# MISSIONS by the NUMBERS

## An Account of 187 Bombing Missions Over Europe (1944-1945)

### Edited by

### Sammy Schneider

*Missions by the Numbers,* 2nd Edition
Copyright © 2000, 2008
Edited by: Sammy Schneider


ISBN: 978-0-9713538-3-1


Published by
**Tarnaby Books**
2576 Fox Circle
Walnut Creek, CA. 94596

EAJWWhiting@aol.com


PRINTED IN THE UNITED STATES OF AMERICA

# Table of Contents

| | |
|---|---:|
| Acknowledgements | 1 |
| Forward - by Lynn Cotterman, Navigator | 4 |
| Preface - by Sammy Schneider, Tail Gunner | 5 |
| Short History - by Earl Bundy, Line Chief | 6 |
| Special Order of the Day - Field Marshal H.R. Alexander | 8 |
| To Honor the 485th Bomb Group - Distinguished Unit Citation Summary | 9 |
| "Destiny"- a poem by John di Russo, Gunner | 10 |
| Introduction - by Lynn Cotterman, Navigator | 11 |
| What is a Mission? - by Sammy Schneider, Tail Gunner | 12 |
| | |
| **Bomb Group Data** | |
|     Briefing Map - Rosenheim Germany, April 19, 1945 | 13 |
|     The Payload-bomb and bombing information | 14 |
|     Flak-Lethal Black Puffs | 15 |
|     Flak photos | 16 |
|     The Bomb Run | 17 |
|     15th Air Force Statistical Summary | 18 |
|     Prisoners of War-Author unknown | 19 |
|     Escort Planes | 20 |
|     485th Bomb Group Statistics | 21 |
|     The "Flimsy" | 22 |
| | |
| Mission Planning - by Master Sgt. Bob Benson, Headquarters | 23 |
| The Crew Chief - by Jess Akin, Crew Chief | 24 |
| The Final Day - by Shirley "Hank" Hancock, Gunner | 26 |
| Abbreviations Used | 29 |
| 485th BG Mission Summaries - Missions 1-187 (10 May 1944 - 25 April 1945) | 30 |
| Perfect Pass - by Jo Haden, daughter of Oliver Haden, Navigator | 137 |
| Formation Position Chart for final 485th BG mission (25 April 1945) | 178 |
| Abbreviations used in Mission Summaries | 179 |
| Conclusion - by Lynn Cotterman | 180 |
| Photos of damaged aircraft | 181 |
| Mission List | 183 |

Dedicated to all the men of the 485th Bomb Group (H) who fought, served and died in defense of their country during World War II.

# ACKNOWLEDGEMENTS 2008
## SECOND EDITION

We have received numerous requests for copies of this book, which has been out of print for several years. As a result of this renewed interest, the decision was made to publish a limited number of copies.

This is NOT a revised edition, although it has a new look and redesigned covers to differentiate it from the original edition. The information contained inside the book is the same as in the original, with a few extremely minor changes in photos and text, for clarity. Although several of those who helped produce the original edition of this book are no longer with us, their efforts are not forgotten and their words live on. Special thanks go out to Carl Gigowski, the first 485th BG Association Historian, whose research produced the mission summaries, to Sammy Schneider, the second 485th Historian, who collected the stories and photos and supplemented Carl's work, and to those who supplied the stories.

The 485th Bomb Group Association gratefully acknowledges the assistance provided by others in the making of this book. Numerous photos and charts were supplied by others, some of them unknown. Most of the photos are 485th Bomb Group photos, many taken by the 485th Bomb Group Photography Unit, but a few were obtained from unknown sources, primarily associated with the 15th Air Force.

The careful reader will note a few errors and omissions in the text. Some of these are a result of Air Force records not being accessible prior to the first printing of this book. Although additional and corrected information has become available since the first printing of this book, through the Freedom of Information Act, the text in the chapters and mission summaries remains as in the original edition.

Our thanks go to Jerry Whiting, the current 485th Historian, who was the driving force behind the publishing of this second edition. Jerry designed the front and back covers, reworked the photographs and consulted with the printer to p roduce an attractive and durable second edition. Thanks, Jerry.

We hope that you, the reader, will enjoy this book.

# ACKNOWLEDGEMENTS

## Sound Off - Sound Off - 1,2 - 3,4

Happy are we that can say another book about the 485$^{th}$ Bomb Group (H) is done. Out of the way-Finito. But then there is always a BUT-that the many who helped in so many different ways, by phone, writing, and verbally with words of encouragement. So, our opportunity has come to do just that-to recognize those wonderful people by naming them.

Taking Off-Thanks to my families in Denver and Rochester, N.Y. I wasn't going to mention the names in my families, but have to mention my youngest grandson, Kevin, in Highlands Ranch, Colorado. He scanned and sent the first picture of the Insignia Flag to Dick Mattison, my crew member on *The Lady* for possible use in the book. Next is a must-Lynn Cotterman, 831$^{st}$ Navigator and Treasurer of the 485$^{th}$. He came by like an angel from Heaven making it known (by phone calls, e-mail messages and 6-hr drives from Albuquerque, New Mexico) that he would help in any way required to finish the book. Without Lynn's enormous contribution and Jo Haden's constant encouragement, this book would not have been possible at this time. Warren Gorman and his wife, Evelyn, were always there for any help when needed. Warren flew in *LIFE* With his crew, Bacon's Beavers. He constantly encouraged me, knowing my physical problems, not forgetting to send me lots of goodies that he made on his computer, which could be used in the book, to make me smile.

If I carry on this way, I'll be writing another book; yet, I must mention their names, Mavis and Jess Akin-two more wonderful helpers. This time, to use a picture of *LIFE* on the cover, as well as a story by Jess about the ground crew. We received a letter from Dr. Edward Alderman Scott, PHD, Director of the Air Force Academy Libraries. It's my sincere thanks to him, giving us permission to use the picture of LIFE. Mavis also donated 3 other paintings to the Academy, depicting planes over a target or enroute home after flying through those black puffs (AKA FLAK). Her kindness gives the 485$^{th}$ publicity for the many who see her paintings reminding them what WWII was all about, and that it gave us the FREEDOM WE HAVE TODAY.

Thanks to Shirley W. Hancock for telling about his experience on the mission to Blechhammer, Germany, when Walter E. (POP) Arnold, (our first Group Commander and Major General, ret today) and crew were shot down, showing us what war is all about. Other stories as well; authors unknown, that helped make this book more interesting reading; Earl Bundy, our former Coordinator, for his short history of the 485$^{th}$, that he wrote for the dedication of the plaque at the Memorial Wall of the Air Force Academy, September 29, 1987. We used it again and we say thank you. I have to say special thanks to my Mystery Woman, who I met via the internet, Jo Haden Galbraith. Jo is another Angel who has helped in so many ways. Her queries came to me regarding her father, Captain Robert (Bob) O. Haden, 831$^{st}$ navigator.

After some research with the records I have (passed down to me by Carl Gigowski), I was able to answer her many questions. She had always wanted to know something more about her father's war record. I have done this for many others over a period of years and it always makes me a happy camper.

Speaking about Jo, she has also helped with proof reading and so she wanted to express her immense gratitude to Sammy (Schneider) and Lynn Cotterman for their expert technical assistance regarding her father's mission story, and for their unwavering friendship. She extends a special thank you to all the men of the 485th Bomb Group who willingly shared their stories and memories with her. Lynn and I accept her acknowledgement. It feels nice to be recognized from time to time, don't you think?

We must not forget Bob and Lynda Hanson-LWT editors and our reporters, who keep the group advised of what's going on by sending our Lightweight Tower newsletter- Warren Sortomme, Headquarters; Sherrill Burba, 828th; Joe Cathcart (Journeys End), 829th; George Dyer, 830th; Lynn Cotterman, 831st, Sammy Schneider, Historian, 485th Bomb Group, and not to forget Mike Kilbury, who puts out all the 485th rosters.

Another edition not to forget, namely Kinko's, your one-stop printing shop who took on the printing of our book. Thank you Cindy Wiberg for getting all the changes made so our mission is completed. There are so many more names to mention and I am at a loss to know where to begin.

**Sammy Schneider (1916- 2006)**
**485th Bomb Group Assoc. Historian**

## FOREWORD

DECEMBER 7th PEARL HARBOR! The freedom of our country was threatened. We received "Greetings" * and answered the call much the same as our forefathers who were called from their homes to fight for Independence. Now we were called from our communities to defend it. We were young men from different backgrounds. We subscribed to different beliefs and life styles. Some of us came from big cities and were "street smart" while others came from small towns and rural areas and were naive. However, none of us had any idea how much our lives were going to change.

We were a typical bunch of young Americans who formed the 485th Bomb Group along with a few career servicemen. After intensive training we were assigned to fly combat missions over Europe from Venosa Air Field in Italy. We developed into a team. Each man had a job to perform and our success depended on each doing his job well, both the men on the ground and the men in the air. A great comradeship developed; a relationship that is still with us today.

Sometimes the youthfulness was displayed among the aircrews. They could party and be frivolous and apparently irresponsible on the ground, yet utterly close net and professional in the air.

We are proud of our contribution in preserving the freedom of America. Sadly we mourn for our buddies that did not come home. Freedom comes with a price.

By Lynn Cotterman
Navigator 831st Squadron

# PREFACE

Once again, I completed another mission. Let's call it # 00000000, like my crew 6 mascot SMOKEY'S dog tag. I bought him for 2 cartons of Raleigh Cigarettes from an Italian lad who used to deliver plum tomatoes to our crew on request.

Happy are we that this mission is completed. It seems for years, flying personnel wanted to know what happened on mission # so and so. Perhaps a mission they went on. They would remark, it's listed in THIS IS HOW IT WAS & THE HISTORY OF THE 485TH BOMB GROUP, but not narrated. Like a broken record, I would explain that I was given 6-month deadline to finish the History. Please realize that it was an impossible task to narrate, repeat narrate, all 187 missions our 485th Group went on.

The Missions from May 10th, 1944 to bomb the Marshalling Yards in Knin, Yugoslavia to April 25th, 1945, another Marshaling Yard in Linz, Austria. Had I been given another 6 months (which the printer took to print the book-TIHIW), all the missions would have been printed as they are here.

And so MISSION COMPLETED, but again credit must be given to CARL GIGOWSKI, who compiled all these narrations. He was our former Historian who died of Leukemia in 1993

Sammy Schneider
Tail Gunner
828 Squadron Crew 6

# A SHORT HISTORY OF THE 485TH

The 485th Bomb Group (Heavy) was constituted on the 14th Sept. 1943 at Fairmont Army Airfield, Boise, Idaho. Activation of the 829th Bomb Squadron was accomplished on the 28th of September 1943, with a nucleus of personnel drawn from the 43rd Bomb Squadron, stationed at Gowen Field. The 11th Anti-Submarine Squadron was redesignated the 831st Squadron and assigned to the 485th Bomb Group on the 20th September 1943. The 15th Bombardment Operational Training Wing Replacement Pool, at Gowen Field, was the main source of combat crews received, with additional crews transferred to the group from the Army Air Field at Pocatello, Idaho.

Col. Walter E Arnold Jr. became the first Group Commander and the Squadron leaders under him were

Capt. Edward H. Neft . ........ 828th Squadron CO
Capt. Maurice W. Boney .... 829th Squadron CO.
Capt. Richard V. Griffin ...... 830th Squadron CO.
Capt. Daniel L. Sjodin ........ 831th Squadron CO

Departing Gowen Field, for Orlando, Florida on the 28th September 1943, the basic Air Echelon trained for 30 days in an endeavor to meet future combat conditions. Returning to Fairmont, we continued training of AirCrews and Ground Echelon 24 hours per day, until completion, March 14, 1944.

    The combat crews proceeded to Lincoln, Nebraska, for final procession prior to overseas deployment on March 14th to Morrison Field, Florida. Leaving Morrison Field individually, the Bombers crossed the Atlantic via the Southern Route arriving in Oudna, Tunisia, on March 18th. For the next month, the combat crews engaged in tactical combat exercises. All of the Bombers, and Crews arrived at Venosa Air Field by the 22nd of April.

    The Ground Echelon departed Fairmont, Nebraska. On the 11th of March, for Hampton Roads, Virginia - Port of Embarkation. The Ground Echelon began their voyage to Italy on the 2nd of April. The crossing of the Atlantic was uneventful until the 20th of April as the convoy approached Cape Bengut, near Algiers, where the German Luftwaffe attacked them. They sunk the D. E. Lansdale and 2 merchant vessels. On board one of the merchant vessels, The US Paul Hamilton, were 154 from the 831st Bomb Squadron. We lost almost all of the ground personnel in the 831st.

    By the 30th of April all personnel and aircraft were at Venosa Air Field. On May 10th the Group flew their first mission to Knin, Yugoslavia (Marshalling Yard), and completed 16 missions during the month with the loss of three aircraft. The peak number of encounters with enemy aircraft was reached in June, when 481 attacked our formation. There was a slight decline in July and a pronounced decline in August.

November marked the beginning of winter season and MUD. The weather continued to deteriorate during the following months, restricting the number of missions. On the 8th of February 1945, the Group was awarded the Distinguished Unit Citation, for a highly successful mission to the Florisdorf Oil Refinery in Vienna, Austria on 26th of June 1944.

    A grand total of 187 missions were flown from the period of the 10th of May 1944 to the 25th of April 1945 - dropping 10,550 tons of bombs on enemy installations. During the month of May 1945, the Group returned to the United States and was deactivated on the 4th August 1946.

During its combat tour the Group participated in the following campaigns: NORTHERN FRANCE, SOUTHERN FRANCE, AIR OFFENSIVE-EUROPE, ROME-ARNO, RHINELAND, PO VALLEY, NORTH APENNINES, NORMANDY AND AIR COMBAT EAME THEATRE

By: Earl Bundy
    829th Squadron

ALLIED FORCE HEADQUARTERS
April, 1945

## SPECIAL ORDER OF THE DAY

### Soldiers, Sailors and Airmen of the Allied Forces
### In the Mediterranean Theatre

Final victory is near. The German Forces are now very groggy and only need one mighty punch to knock them out for good. The moment has now come for us to take the field for the last battle which will end the war in Europe. You know what our comrades in the West and in the East are doing on the battlefields. It is now our turn to play our decisive part. It will not be a walk-over; a mortally wounded beast can still be very dangerous. You must be prepared for a hard and bitter fight: but the end is quite certain - there is not the slightest shadow of doubt about that. You, who have won every battle you have fought, are going to win this last one.

Forward then into battle with confidence, faith and -determination to see it through to the end. Godspeed and good luck to you all.

*H.R. Alexander*

*Field-Marshal,*
*Supreme Allied Commander,*
*Mediterranean Theatre.*

## TO HONOR

### THE 485th BOMB GROUP

The 485th Bomb Group (H), 15th Air Force, in a formation of 36 B-24 aircraft flew a mission to Vienna, Austria on 26 June 1944 to attack the Florisdorf Oil Refinery. The Group inflicted grave and massive damage to the refinery despite heavy flak and intense fighter opposition, and crippled the enemy's vital fuel production during a crucial period of WWII. The 485th Group was awarded a Unit Citation for this successfully completed mission.

"A PERFORMANCE ABOVE AND BEYOND EXPECTATIONS"

By: Veterans of the 485th Group and Walter E. "Pop" Arnold USAF (Ret.), Commander

## "DESTINY"

From The USS Arizona
Pearl Harbor, USA

I was just a farm boy from Iowa
  Who joined the Navy to see the world.
And after some basic training
  They shipped me overseas to Pearl.

I was assigned to the Arizona
  And I was as happy as I could be.
To me it was a beautiful ship
  So I named it, 'KING OF *THE SEA*'.

I trained as a Gunners mate
  And enjoyed it every minute.
It made me proud and I just knew
  If we got in a war the ship would help
us win it.

So on a pretty Saturday afternoon
  December 6th, 1941 to be specific
We took a shore leave and,
  Hit every bar in the South Pacific.

Soon our leave was over
And we all headed on back
Looking forward to Sunday morning
When we got extra time in the sack

Before going to sleep that night
  We talked aid kidded about life after Pearl
About the plans, we had and
  Making good in a civilian world.

LIGHTS OUT! SLEEP TIGHT!
  Was blared out over the horn.
The weatherman says it's going to be
  A beautiful Sunday morn

AND SO WE WENT TO SLEEP.

What happened next is very, very hard to believe.
  An action took place that only barbarians could conceive.

The Arizona took one hell of a hit,
  A Torpedo down the stack,
The ship went down so very fast
We had no chance to even fight back.

Life came to a end quickly, like
  A thief comes in the night
For us it was all over
We would never join in the fight.

Why did it have to happen that way?
Only God in His heaven has the right to say.

We had a lot of love to give.
We had a lot of life to live.
Some things are really just meant to be
And so we fulfilled our "DESTINY"

My love of the ocean
Meant the whole wide world to me.
Now I'll sleep life eternal
in the bosom of this beautiful sea

So when you think- about us
Please pause and say a prayer.
And our souls will rest more peaceful
To know that you really care.

From: Just an Iowa Farm Boy
         Known Only Unto God.

Written by:  John DiRusso
             U.S. Air Corps
             WWII Combat Vet

I call John DiRusso a one man army- Patriotic to the first degree and constantly fighting for Patriotism. The first time he wrote a poem --Please Remember Me, which he fought for & won to get included in the Congressional Record. --2nd he fought for and won, with the Washington Postmaster, to have all our flags flown at half mast on Memorial Day in Post Offices in all our 50 States. & 3rd-he presently fighting to get bill HR965 he had introduced in Congress by Jack Quinn & Senator Arlen Specter and hoping for a YES VOTE so it will make Pearl Harbor a National Holiday.    Sammy Schneidner

# INTRODUCTION

Originally, it was our intent to compile the information of the 187 missions flown by the 485th Bomb Group into a book so each veteran could find the records of the flights that he flew.

After reading through the accounts of several missions it was evident that since the information is mostly facts and figures, the book would not be as interesting to the readers who did not participate in the missions. It was decided to expand the information so our offspring could better understand what was involved in flying a combat mission. Some material that has been included describes some of the nomenclature and procedures of preparing for a mission and the teamwork involved in a successful bomb run.

The first missions that were flown by the 485th were called "Freshman Missions". The targets were marshalling yards in Italy and Yugoslavia to gain combat experience with the least causalities. Some flak was experienced on some of the early missions, but the mission to WIENER-NEUSTAIDT to bomb the airdrome, was the first time they experienced heavy and accurate flak.

The main thrust of the 485th Bomb Group was the bombing of marshalling yards and oil facilities. Other targets included airdromes, industrial areas, communications, ammunition dumps and bridges. In April, the last month of the war, many bridges were destroyed in northern Italy to cut off retreating German troops. Marshalling yards in many cities were bombed depending on the rail traffic. However, attacks of oil refineries were more concentrated. BLECHHAMMER, MOOSBIERBALM, PLOESTI and VIENNA received the most bombardment.

Some people wonder why the same targets were bombed several times. Well, it was not possible to cut off rail traffic for any extended length of time. The enemy could repair the tracks and have a through line running within hours. Consequently the groups of railroad cars that were parked in the marshalling yards were the targets. If the marshalling yards that had been bombed previously, were full of cars again, they went back on the target list. "Repeats" to oil facilities were necessary because the enemy made every effort to repair his oil producing and refining facilities so frequent "policing" missions were required.

With the development of nuclear weapons and the advent of jet power, a chapter in aerial warfare came to a close. The B-24 Liberators became obsolete, but there will always be a soft spot in our hearts for that old bird that could take a lot of punishment and still bring us back to Venosa. She was a reliable partner in the fight for freedom.

By Lynn Cotterman
Navigator, 831st Squadron

## What Is A Mission

The Winston Dictionary states one of a number of meanings - as the sending or state of being sent with certain powers to do some special service. The mission of which we speak was in essence a Mission of War, ordered by the higher echelon to go to a place (one or more) they select; and to devastate with a torrential rain of high explosives namely bombs of many types. We must also remember the terminology of friend or foe because this was a time of WAR. The Grim Reaper is always present as you approach the selected Target for the Day. Havoc is all around you and the Grim Reaper in the form of enemy fighters or antiaircraft guns firing flak whose purpose was to stop you before you can rain bombs on the target. It's touch and go never knowing when the Reaper will reach out and pick you as you fly through the flak, observe planes on fire and exploding and crew members bailing out. If they were lucky they would become Prisoners of War or maybe it's your time to die.

A mission is foreboding from the time you are awaken in the early morning hours from your warm cot to going through the morning ritual of a Latrine stop, washing up, and going to the mess hall. Then it's on to the briefing room and the lowering of the drape showing the wall map pointing to the target for the day. We receive information regarding weather, our fighter planes, P-47's, P-51's or P-38's escorting us to and from the target, the number of anti aircraft guns we could expect and the number of enemy fighters in that area. Then the Chaplin ended the briefing with a prayer and "GOD SPEED WITH A SAFE RETURN"

Next it's on to the equipment shack to pick up the necessary gear you will require for your safe being, like a parachute (seat or chest pack), flak jacket, flak helmet, oxygen mask and heated flying suit. You head for the revetment area where your plane is parked. It has been pre-flighted by our ever so needed ground crew. They see that all is ready for today's mission, with sufficient fuel, oil, oxygen, tire pressures, and of course the deadly target bombs and ammunition for all the 50 caliber machine guns.

Next we wait for the flare which tells us we are ready to go for our mission trip to that TARGET FOR TO-DAY. With our crew on board, the plane engines purring, a plane from our Bomb Group Is airborne every 30 seconds heading to rendezvous with other bomb groups of our 55th Bomb Wing. We are on our way to the target we have to bomb. A mission usually from 3 or 4 hours to 10 hours round trip but each mission can be that one where HAVOC is waiting to do damage to part or all of your plane and the crew.

Sammy Schneider
Tail Gunner
828th Squadron Crew 6

**BRIEFING MAP**

This is a photo of the huge wall map in the briefing room showing the route for mission number 183 to Rosenheim, Germany, 19 April 1945. Compass headings are shown along the route (MH). The dark areas are flak areas surrounding the cities that had potential targets. The long area shaped like a worm is the Brenner Valley that connects Italy with Austria. Each crew was given a photo of the map.

## TYPES OF BOMBS

The Allies had different types of bombs for different purposes. Armor-piercing bombs had thick skins and sharp points that allowed them to break through a ship's hull or a tank's body and explode inside. Anti-personnel or fragmentation bombs had thin skins that exploded into hundreds of steel splinters when dropped on troops. Incendiary or fire bombs were loaded with chemicals that burned at high temperatures and usually weighed less than one hundred pounds apiece. Demolition bombs were the most common; they were packed with high explosives such as TNT or Amatol and were used to blow up bridges and structures and other things.

## BOMBING STRATEGIES

Three main classifications developed: battlefield close support; medium-range tactical strikes, such as bridges, trains, troop support areas and ammo dumps; and lastly, "strategic" bombing of "military" targets. An example of the last is aircraft factories making enemy airplanes; hence, this term came to mean the bombing of both "military and indistrial targets".

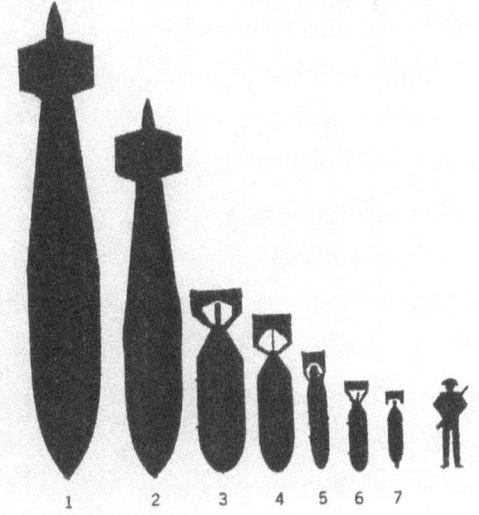

*Some British and American Bombs of World War II*

1. RAF 22,000-pound Grand Slam, the largest high-explosive bomb ever made
2. RAF 12,00-pound Tallboy, used for breaking through the concrete roofs of U-boat pens
3. USAAF 4000-pound Blockbuster
4. USAAF 2000-pound Blockbuster
5. USAAF 1000-pound armor-piercing bomb
6. USAAF 500-pound standard demolition bomb
7. USAAF 100-pound bomb with nose fuse; shell could also be filled with white phosporous to make the "Kenney Cocktail"

## UNIDENTIFIED B-24 DROPPING ITS HEAVY BOMB LOAD

*A battery of German 88-millimeter fluk guns.*

## FLAK - LETHAL BLACK PUFFS

Flak" is a shortening of (Fl)ieger(a)bwehr(k)anone, German for antiaircraft gun". There were two types of flak over German. Light flak, .20 mm, was dangerous up to 20,000 feet. Heavy flak, .88 mm cannons, could hurl a shell three inches in diameter as high as 40,00 feet. Some flak shells had clock mechanisms times to explode into wing tearing, fuselage-ripping fragments. Others had fuses so sensitive that the slightest touch of a wing would set them off. Antiaircraft shells were so powerful that any plane within 50 feet of an explosion was sure to be hit by high-speed steel fragments.

As we approached the target area we could see the black smoke-like puffs forming a large polka-dot pattern that we knew we were going to fly through. Sometimes the amount of flak that a box (six or seven planes) received was the luck of the draw. A barrage might miss one box and the ones behind it would catch hell. Sometimes everyone caught hell. We soon learned how lethal those little black puffs were. A single burst could rip off a wing in a second and send a plane into a lazy spin. In fact flak was more deadly than most of us knew. In 1944 German flak destroyed 3501 American planes; nearly 600 more than Lufwaffe fighters!

Sometimes we would watch those black puffs spit and wonder how the Germans could continue to miss us. A gunner said that flak was his nightmare. One could always fight back with fighter planes, but not flak: all one could do was sit and pray that an 88 wouldn't hit your ship and blow it into small pieces. It didn't take us long to figure out that the more missions that we flew, the probability of taking a direct hit increased.

**FLAK WAS LETHAL**

**55TH BOMB WING LIBERATOR WINGS ITS WAY THROUGH FLAK OVER A EUROPEAN TARGET**

**THE BOMBERS HAD NO OTHER CHOICE, BUT TO STAY ON COURSE AND IN FORMATION AND FLY THROUGH IT**

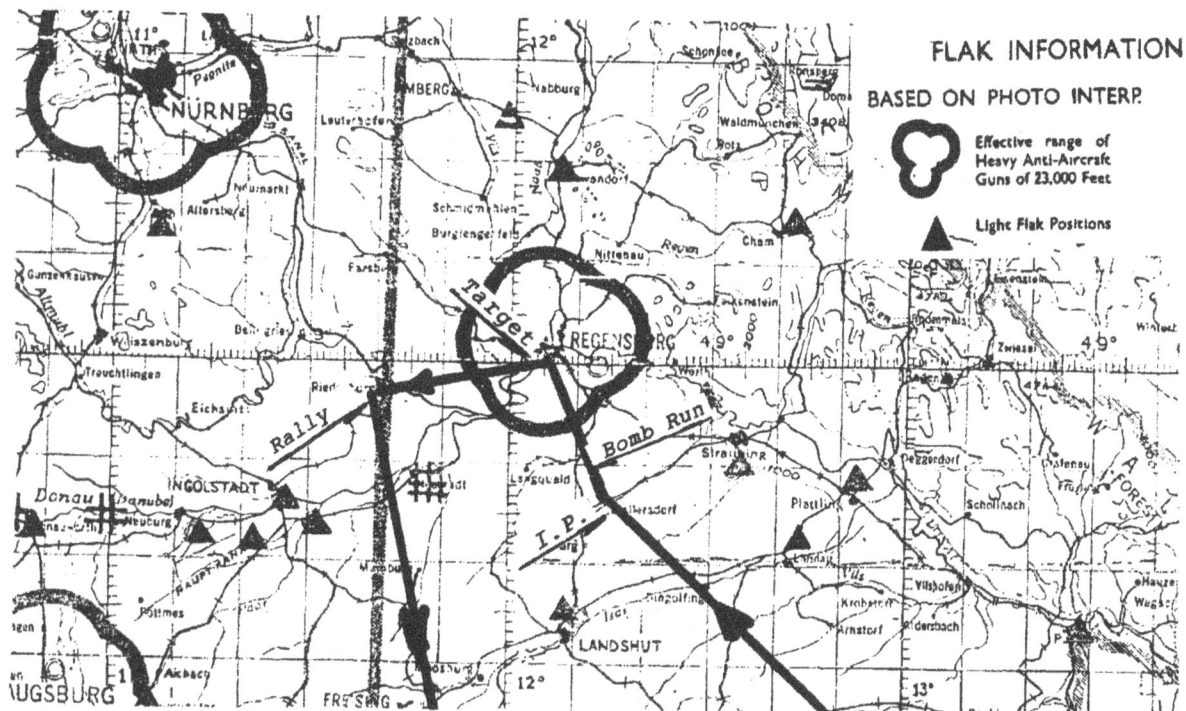

## THE BOMB RUN

The bomb run began when the group turned toward the target at the Initial Point (IP). The IP was selected outside the flak area and usually 10 to 15 minutes of time before "bombs away". Sometimes the IP would be beyond the target and we would turn back to the target hoping to catch the enemy off guard, but I'm not sure that this was effective.

After the turn was completed we entered the flak area. The bomb bay doors were opened and the formation was tightened up to get a good bomb pattern. The success of the mission now depended on the skill of the lead bombardier. The pilots held the formation steady so the bombardier could get a bead on the target. This was the moment that culminated all the training of the air crews and the skill of the ground crews that prepared and armed the planes. Since there were no fighters in the flak area the gunners had to sit and hope an .88 didn't blow us up. The Lead Bombardier made the final corrections and locked in the Norden Bomb Sight.

"Bombs away". The bombardiers dropped the pay load. The Navigator recorded the time, temperature, heading, air speed and altitude to help in the critique of the mission. The Armor/Gunner who was in the ball turret called to report if any bombs had hung up and had to be released by salvo. The group usually made a sharp turn to leave the flak area as soon as possible. Then the bomb bay doors were closed. They had been left open in the flak area as an escape route. The groups gathered at the rally point where they reformed back into a large formation. The groups did not go over the target all at one time, but in a "follow the leader" style. The next moment was an anxious one as everyone checked in to make sure no one was injured and that they were still on oxygen. The pilots assesed the damage and checked the amount of fuel left as we headed back south. We weren't out of enemy territory yet, but we tentatively marked off one more mission.

## STATISTICAL SUMMARY

### B-24 Gains
| | |
|---|---:|
| From XII BC | 90 |
| New | 809 |
| Replacement | 2,547 |
| Other | 98 |
| Total | 3,544 |

### B-24 Losses
| | |
|---|---:|
| On Hand 10 May 1945 | 1,067 |
| Combat | 1,756 |
| Non-Ops | |
|   Accidents | 192 |
|   War Weary | 281 |
|   Other | 248 |
| Total | 3,544 |

Sorties: B-17s and B-24s in the Fifteenth Air Force averaged 8.5 sorties per aircraft per month.

### B-24 Aborts
| Cause | Number | Rate |
|---|---:|---:|
| Weather | 15,255 | 13.5% |
| Mechanical | 5,511 | 4.9% |
| A/C and Accidents | 2,481 | 2.2% |
| Lead Ship Fail | 831 | 0.7% |
| Miscellaneous | 2,302 | 2.0% |
| Total | 26,380 | 23.3% |

### B-24 Sortie Record
| | |
|---|---:|
| Airborne | 113,218 |
| Early Returns | 10,196 |
| Total Sorties | 103,022 |
| Non-Effective Sorties | 16,180 |
| Effective Sorties | 86,838 |
| Percentage Effective Sorties | 84% |

### B-24 Victories and Losses
| | |
|---|---:|
| Encounters | 8,703 |
| Destroyed | 1,274 |
| Probable | 505 |
| Damaged | 336 |
| Victories | 2,115 |
| Flak | 803 |
| Enemy Fighters | 377 |
| Mechanical | 313 |
| Unknown | 257 |
| Losses | 1,750 |
| Damaged Category I | 7,583 |
| Damaged Category II | 13,346 |
| Total | 20,929 |

### Casualties
| | |
|---|---:|
| Killed in Action | 2,703 |
| Missing in Action | 12,359 |
| Wounded in Action | 2,553 |
| Total | 17,615 |

### Effectiveness Data

Bomb Tonnage:
- 90,914 Italy
- 74,211 Austria
- 35,927 Germany
- 26,364 Rumania
- 76,426 Europe (other) & Africa
- 303,842 tons on enemy targets in 12 countries, including 8 capital cities.

Sorties:
- 148,955 Bomber
- 87,732 Fighter
- 236,687 80% of total effectiveness

**Aircraft Lost:** 2,380 Bombers (1,756 B-24s)
924 Fighters
3,304 Total

**Aircraft Damaged:** 14,000

**Enemy Aircraft Destroyed:** 1,946

**Enemy Transport Destroyed:** 1,600 Locomotives
1,400 Rail cars
800 Motor transport

# PRISONERS OF WAR

Over 75% of the prisoners of war held by the Germans were aircrew members, who had been shot down over enemy territory. Many attempts to escape were made, but only a small number got back to friendly territory. For the majority, freedom came as the victorious allied armies moved across Germany.

Among the first acts of all prisoners was to communicate with their families. The War Department, as a matter of course, sent to the next of kin, the rather chilling telegram with the words 'MISSING IN ACTION' in it. The International Red Cross at Geneva, Switzerland, served as the clearinghouse, for clearing up " the status of personnel". Also the Germans issued a Post Card with which the prisoner could communicate with his family as soon as he arrived at a permanent camp.

As soon as it was known that a man was a prisoner of war in Germany, the office of the Provost Marshal immediately informed the next of kin, how they might communicate with the prisoner and how they might send parcels.

One of the prized issues, given the prisoners by The War Prisoners Aid of the YMCA was "A Wartime Log". Into this chunky volume with the rough tan cover, the men noted their thoughts, dreams, hopes and remembrances of times past.

The prisoners also made drawings in their logs, cartoons, and portraits of fellow prisoners, maps on which they followed the course of the war. Someone always managed to have a radio, which was tuned into BBC, and then the news disseminated via a camp newspaper POW. The men drew the floor plans of their homes, made lists of their menu in camp and the ones they devour Men they returned to their homes

What happened to a Liberator Crew that had to bail out over enemy territory? Those who were neither captured nor killed (by incensed civilians) sometimes managed to make it back to Venosa over various routes. Most were captured and spent the rest of the war in a German prison camp as KREEGSGEFANGENEN, prisoners of war-or as they themselves preferred, KREEGUES". Then began a lift, of almost unremitting boredom, punctuated by escape attempts and other events of interest.

To keep themselves occupied, besides plotting escapes, the men organized games, clubs, study groups, discussion groups, and libraries, even built models of their favorite aircraft. They also formed orchestras, glee clubs and produced musicals, shows.

STALUG LUFT I was a special prison for fliers. The LUFTWAFFE, as a rule treated their fellow airmen better than the other prisoners of war handled in other camps.

Author Unknown

# "Little Friends"

15th Air Force P-38 Lightning fighters in nice formation shot.

Two of the 82nd Fighter Group P-38 pilots, Chester "Ike" Eckermann and William "Pat" Patterson, who protected our 15th Air Force bombers from fighter attacks.

One of our escorts, a P-51 Mustang fighter from the 325th Fighter Group, the "Checkertails", lands at Venosa for a visit.

## ESCORT PLANES

We were escorted on each mission by P-38's, P-47's or P-51 fighter planes to defend against the Luftwaffe. We made a rendezvous with them, usually just before we entered enemy territory. They stayed with us until we reached our target and then another group replaced them and escorted us back to friendly territory. They were always a welcome sight.

# STATISTICS

485th Bomb Group
828th, 829th, 830th, 831st
PROJECT NUMBER - 90526
Phase Training - Fairmount AFLD, NB - September 20, 1943 thru March 11, 1944.
Trained in Tunisia during ~March-April-1, 1944.
Venosa, Italy - April 30, 1944 thru May 9, 1945
1 DUC - June 26, 1944.

AIRCRAFT SUMMARY

|  | Squadrons | | | | | |
|---|---|---|---|---|---|---|
|  | 828th | 829th | 830th | 831st | UNK-BS | TOTAL |
| COMBAT LOSSES | 14 | 19 | 14 | 12 | 0 | 59 |
| OPERATIONAL LOSSES, INCLUDING CON. ANS SAL. | 20 | 13 | 10 | 10 | 9 | 62 |
| KNOWN TRANSFERS | 5 | 1 | 0 | 0 | 0 | 6 |
| ZI-HOME RUN | 10 | 8 | 8 | 13 | 7 | 46 |
| SURVIVED WAR (SAL. AFTER WAR'S END) | 4 | 3 | 4 | 7 | 3 | 21 |
|  | 53 | 44 | 36 | 42 | 19 | 194 |

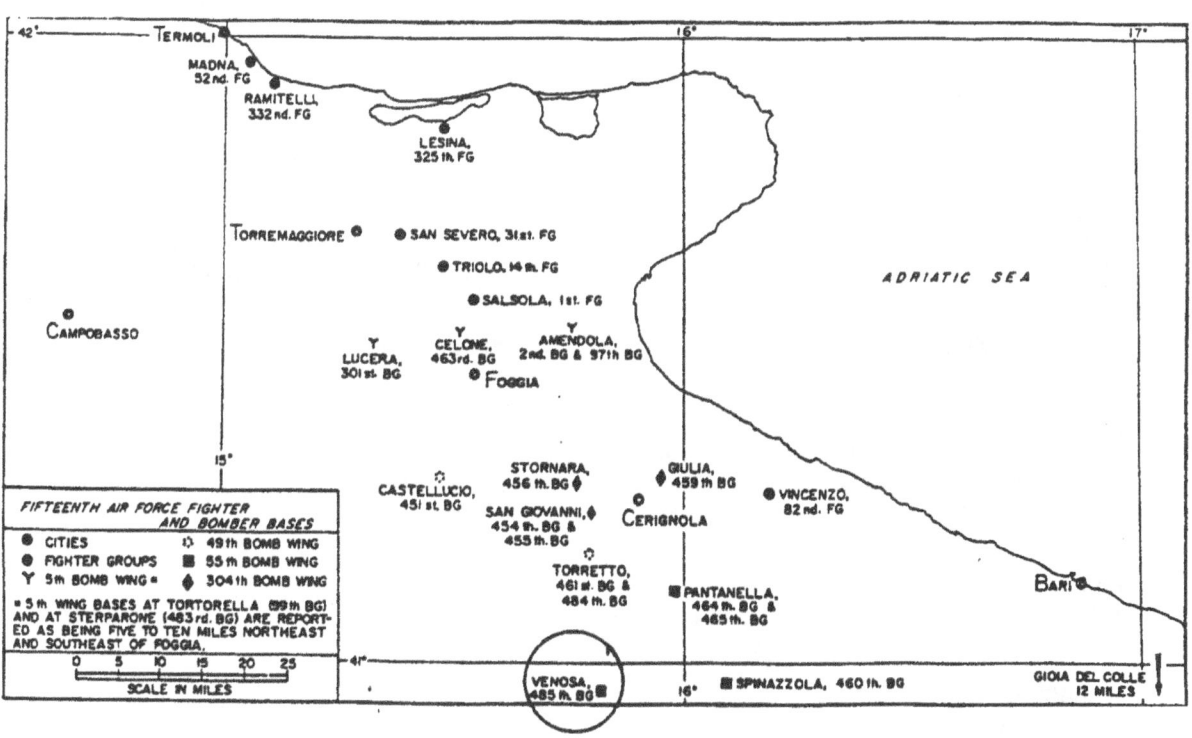

15th AIR FORCE FIGHTER AND BOMBER BASES

SECRET                                                                                        SECRET

## PILOTS FLIMSEY for 20 APRIL 1945

| | | | | | |
|---|---|---|---|---|---|
| Crew inspections | 0735 | Stations.- 0745 | | Start. Engines | 0755 |
| Taxi | 0805 | Take off:: 0815 | | Last possible take Otf | 0900 |

**GROUP ASSEMBLY**     over home field at 7000'

**WAVE RENDEZVOUS**     485th leads 460th over Spinazzola at 0942. 465th & 464th precede 485t!h & 460th 15 minutes. Each two groups act as a combat wave.

**ORDER OF FLIGHT**     Right column - 5th (3 Gps), 47th Wings
Left columrin - 304th, 5th (3 Gps), 49th, 55th(465-464-485-460

**FIGHTER ESCORT**     24 P-51's of -332nd Fighter Group meet -55th Wing. at PISTORIA at 1118 at 20, 000' and provide general. escort.

**KEY POINT**     PISTOIA (4356 1055) at 1133. at 20, O00'

**START CLIMB**     to. Bombing altitude AT SPINAZZOLA

**PRIMARY TARGET**     GARZARE ROAD BRIDGE, (Visual) at 1205
**INITIAL POINTI**     BONDENO (4453 1126)
**AXIS of ATTACK**     33 Degrees
**METHOD of ATTACK**     By boxes, in very close column
**TARGET ELEVATION**     25'
**BOMBING ALTITUDE**

| | | | |
|---|---|---|---|
| "A" box - 21,500' | "D" box - 21,000' | 465th | 21,000' |
| "B" box - 21,300' | "E" box - 20,800' | 464th | 22,500' |
| "C" box - 21100' | "F" box - 20,600' | 460th | 23.,000' |

**RALLY**     Left to MONTAGNANO (4514 1128)-

**FIRST ALTERNATE**     AF #5    VERONA PARONA RR BRIDGE (Visual only)
**INITIAL POINT** :     POSINA (4547 1115)
**AXIS Of ATTACK** :     215 Degrees
**METHOD of ATTACK**     Same as Primary
**RALLY**     Right to DESENZANO (4528`1042)
**ROUTE BACK**     DESENZANO to PISTOIA to BASE

**TARGETS of OPPORTUNITYt**

| | | | | | | |
|---|---|---|---|---|---|---|
| AF # 7 | ORA RR BRIDGE | Visual Only | IP | NATURNO | 4628 1100 | RALLY Left |
| AF #8 | BRONZOLO M/Y | Visual Only | | CASTELEBELLO | 4637 1054 | Left |
| AF #10 | FORTEZZA/AICA Right | Visual only | | MAYRHOFEN | | 4710 1152 |
| AF #11 | BOLZANO M/Y | Visual only | | ST LEONARD | 4638 1114 | Left |

**SPECIAL INSTRUCTIONS.**
**GAS LOAD**     2700,gallons.
**BOMBS**     5 1000# RDX fused .1 nose and .1 tail
Any A,/C bringing in seriously wounded land at BARI or FOGGIA\ MAIN
Any AIC bringing in slightly wounded land at IESI or PANTANELLA

5th Wing attacks VIPITENO & FORTEZZA M/Y at 1042 to 1210
47th Wing attacks NAIRHOF RR VIADUCT
49th Wing attacks LUSiA. ROAD BRIDGE at 1130 to 1145
304th Wing attacks CAMPODAZZO & PONTE GARDENA RR BRIDGE at 1115 to 1135

INDIVIDUAL BOXES TAKE EVASIVE ACTION ON RALLY
ATTACK ONLY TARGETS SPECIFIED ON THIS FLIMSEY
TAKE EVERY PRECAUTION TO AVOID DROPPING BOMBS ON OR NEAR BONDENO
DUE TO A POW CAMP LOCATED THERE

# MISSION PLANNING
## GROUP HEADQUARTERS OPERATIONS

Group Operations (S-3) was comprised of officers and enlisted men - each specializing in either navigation, bombing, radar, weather, aerial photo interpretation, armament, gunnery, or communications. Group Operations was headed-up by the Operations Officer and the Master Sergeant. A warning system for hostile aircraft was also operated by Group Operations.

Prior to the early morning briefing, there was a great amount of planning and preparation by many men in Group Headquarters and in each Squadron. It began the previous afternoon when Group Operations received a coded message from Wing Headquarters. That message stated the number of aircraft and bomb load needed for the next day's mission. This information was passed on to Group Engineering, Armament and Intelligence and to each Squadron. The target and route followed later. The men in these units began work on their assigned duties in preparation for the mission. Meanwhile on the flight line the planes were getting their final check for readiness and the bombs were being loaded into the bomb bays.

Along about 10:00 or 11:00 o'clock in the evening, two men from Group Operations were dispatched to meet a courier at a pre-designated location to receive the sealed orders for the next day's mission. Upon returning to the base the Group Commander, his Deputy and the operations Officer reviewed the orders and the "Flimsy" * and every man went to work through the night on their respective duties for the morning briefing. The object was to make sure each crew had the latest and most complete information so they would have the best chance of a successful mission and a safe return. The route was reviewed and drawn on the huge briefing wall map. Information about the target was collected for the briefing; photos of the target were studied, the weather was forecast, the latest information about the fighter opposition and flak at the target was gathered, etc. All this information was duplicated so each crew would have a copy. The Group Commander designated who would fly the lead position. There was constant communication between Group Operations and each squadron headquarters to assure the readiness for the mission.

It was always hoped that all aircraft and crews would return safely to home base, but this was seldom the case. Following the mission each squadron Operations Officer reported to Group Operations the number of aircraft lost. It was a great day when all aircraft and crews returned without a mishap and we prayed that this would occur on every mission.

Bob Benson, M/Sgt., Group Operations

* The flimsies contained all the detail information about the mission. They were called flimsies because in the beginning they were made from rice paper and could be eaten quickly if in danger of capture. The flimsy for the mission on April 20, 1945 can be found under the heading "DATA. One can see the amount of detail involved. Nothing was left to chance.

# THE CREW CHIEF

## JESS AKIN

I was an assistant crew chief to Lloyd Arnold. After 30 or 40 missions our B-24 burned to the ground on the hard stand. An inquiry was held and Lloyd was cleared and was sent to Rome for R & R. After only 3 or 4 minutes in the air, his plane with all its passengers including Lloyd Arnold crashed and they were killed. So I moved up to crew chief.

In the beginning we had no personal because most of the 831st ground crew was lost when the Liberty ship, Paul Hamilton, was sunk on the trip over. We borrowed personnel from other squadrons, which left all of us short handed. A Liberator should have a Crew Chief, an assistant crew chief and two mechanics to do all the work on a B-24. We ended up with two men to a plane; some had three. So you and your assistant had to pre-flight by yourselves. All four props had to be rotated four times before you started the pre-flight. This was necessary to get the oil up to all engines. In freezing weather, that was one of the toughest jobs for only one man to rotate the props by hand, at least four times.

A flight chief was in charge of four planes. He in turn reported to the Line Chief and he reported to the Engineering Officer. Basically, on the line, there were engineering, armament, communications, ordnance, prop specialists, radio, sheet metal specialists, electricians, bombsight technicians, power turret specialists, regulars and inspectors. The maintenance mechanics had to keep a B-24 in flying condition so that it would take its crew to the target and home safely.

The Crew Chief and his assistant were in complete charge of the plane. After a mission, the Pilot would report all damages to your plane; you and your crew would give it a good going over; and see if it could be fixed for the next flight. If not, you put it on a RED X and told the Line Chief so he could report to Group Operation the number of planes that would be ready to go the next day. A RED X meant, fuel tank leaking, brakes out or the engines shot up. Maybe there were too many flak holes and the sheet metal crews were put to work, or there might be electrical trouble.

Besides the mechanical work you had a lot of paper work on all four engines. Every engine had a special sheet to report all the work completed; at 50 hours, an oil change; at 100 hours, all spark plugs had to be replaced, etc.

The days started way before dawn-go to the mess hall for a quick coffee and dehydrated eggs, potatoes, you name it and that is what we ate-SPAM. Then jump on a waiting truck and go to the flight line to pre-flight your B-24. Check everything, oxygen gauges for all ten crew members, etc. then top your fuel tanks and then wait for the Crew to show up.
One morning I heard a great bang and thought one of my engines had trouble. I shut everything down and jumped out to see what did happen. Well, the nose Gunner in the hard stand across from me disobeyed orders and was loading his Cal. 50 machine gun and one bullet had gone off and hit the end of my B-24. The nose gunner came over and asked, "What's wrong, Akin I told him that we found a good four or five inch hole in the wing. He begged me not to report it, as he would be in serious trouble. My plane went ahead on its mission; came back and I wrote it up that it happened on the mission.

The reason a Crew Chief or his assistant is always on flight pay is, because he is in the air almost every week or two. If you put your B-24 on a RED X when it comes back from a mission it would miss the next mission. After a repair was finished, a Pilot and the Crew Chief had to test hop it at least one hour before it could go on another mission.

On one test hop, as we were landing, our nose wheel broke off. For a second, I saw something go by, then sparks started coming, and we ended up nose in the runway and the tail about 15 feet in the air. Akin came out the back opening, and hit the runway running (many a time when they had trouble landing, the plane would catch on fire and be completely destroyed). The ground crew gave me a hard time the way I flew out of that B-24.

Orders came from Headquarters that we had to stop hedge hopping (as we did on most of the test hops) The pilot said to me, "Akin, see that large haystack on that pole-You want to buzz it?" And away the haystack went, with the Farmer waving his pitchfork at us-Line Chief would say, "How did all this hay get into this plane?" On another Red X test hop, the Pilot said, "Akin have you ever seen Mt Vesuvius?" I said, "Yes, from the air'. He said, "You are really going to see it now". We really rotated our B-24 around the crater of Mt Vesuvius about four times-hell yes, it was dangerous and crazy.

On one occasion Squadron Commander Col. Ed Nett told me to go and get my parachute, we are going on a flight. The P-38 pilots use to buzz our tents so close that at times some of our tents were blown down. Well, Col. Ed Nett had a close buddy that was in his Cadet Class; so he was going to give him a dose of buzzing. Col. Neft flew as close as he could to their tents changing the pitch of the props (which creates a lot more noise)-then pulled straight up. I didn't believe that the B-24 would stay together. You talk about a roller coaster-well for get it-this was it!!! (Ed said that Jess' eyes were as big as saucers)

We had good times, bad times and sad times, but the most joyful time was VE DAY when they announced the end of the struggle and that we were going home!

Jess Akin, Crew Chief<br>828th Squadron

# FINAL DAY
## Aug. 27, 1944

We were awakened as usual before dawn on August 27, 1944, to prepare for our 24th mission. After a quick breakfast, we went to briefing. When the target was uncovered we saw that it was Blechhammer, Germany, a very long mission, and were told that there were several hundred anti-aircraft guns in the target area and as it was hit just five days before the Germans may have moved more in. The flight to the target was not bad, a few areas where 88s threw a few bursts of flak our way, but no damage. But as we approached the I. P. things changed.

A group ahead of us actually disappeared in the black cloud of flak. I was in the waist position on this mission and the flak closed in as we were on the bomb run. It was heavier than any we had been in before including two trips over Ploesti oil fields. The bursts were heard and felt and the helpless feeling prevailed that we just had to pray we got through and released our bombs on the target. Just as the bombardier said "Bombs Away" we took a direct hit somewhere in the front of the plane (B-24J). A quick burst of flame came back through the plane and it felt like we hit a brick wall. Almost immediately the second shell hit us and the black smoke was blinding. I couldn't see Sgt. Ted Brown, the other waist gunner, only a few feet away.

Something hit me in the chest and knocked me against the side of the plane and to the floor. I landed on top of my parachute, which was on the floor. I looked down at my chest and saw the flak suit was torn but still in place and I was having trouble getting my breath. My oxygen mask was on the side of my face, which I corrected. It was impossible to realize everything that was going on. We were hit a third time all within a few seconds. I saw Ted Brown get to his feet and together we opened the camera hatch. I released my flak vest and put my parachute on the harness snaps as we could see the bodies of the crew in the front of the plane bailing out. Sgt. O.E. Meyer and Sgt. Bill Killian were close to us Sgt. Tony Annie was in the tail gun position. After several members from the front of the plane had left, and there was no response on the intercom, we started bailing out. With the smoke and noise of the badly damaged aircraft we knew that it could explode at any second.

After my chute opened and I got my bearings again, I started counting parachutes. There were five below and behind me and four above and ahead of my position. That made ten chutes in all and I knew we had twelve men on board. It turned out later that two didn't get their chutes open right away and were out of sight. Everyone got out of the burning plane. Any man that ever bailed out over enemy territory knows the indescribable feeling of going down into an unknown country to an unknown fate. Everyone on the ground would be your potential enemy and willing to take your life.

We watched the plane explode as it neared the ground and right into a group of buildings I thought was a small town.
Bail out was above 20,000 feet so it took some time, it seemed forever, to reach the ground. There was lots of activity on the ground, but I was coming down in a heavily wooded area and the landing was of great concern at the moment. My chute was swinging me like a pendulum and I hit the ground going backwards. I hit a jack leg fence going through the woods along a narrow dirt road. The fence was a tough obstacle and I was knocked out by the force of the landing. When things started to clear in my head again, I was alone and my parachute was draped over part of the fence and a waist high bush. I knew if there was a chance to escape,

that parachute had to be made less obvious. Pulling it down and getting it pushed under a small pile of brush was quite a job. I hurt every place, but also knew I had to get to hell out of there. Stripping off my heated suit and consulting a map from my escape kit and laying the small compass on top of the map, I determined where South was and headed out.

My right leg was bleeding from a small gash that was either caused by a chunk of shrapnel or my encounter with the fence, but the worst hurt was my knees and the large bruise on my chest. I consulted my compass frequently and moved south as fast as I could. After some period of time, I came to a large clearing in the woods, which was in the course I wanted to take. After looking carefully about, I decided to cross the clearing as to go around would have consumed valuable time in getting as far from the target area as possible.

I started across at a run but only went a few yards when a man in civilian clothes detached himself from a tree about 75 yards in front of me. He had seen me before I saw him. He had a red band around his arm and looked like he wore a suit and dark hat. He also had a rifle which he brought to his shoulder in such a manner that I knew he was going to shoot. I turned back to the timber as he fired the first shot. I could hear it hit the trees in front of me. He fired twice more, but the last one he wasted as I was back in the shelter of the timber and moving fast. This time I circled the clearing, got my heart out of my throat and kept going south. After about seven hours and some unknown number of miles from where I landed, my luck ran out.

I crossed a small road and a young male on a bicycle came around a bend and saw me. What a howl that little buzzard made. I continued on and within a couple hundred yards, ran out of timber. It was all open country with a German soldier with a rifle in plain sight. I could hear the sound of vehicles, men, and barking dogs behind me. There was no place to go, so I just stood by a tree and waited.

Within a few minutes I was approached by German army personnel, two of which had dogs. There must have been 3 truckloads of them because the trucks were still in the road when they took me back there. Two soldiers walked me down this road a short distance, there made me strip to the skin and examined my clothes. One man in the building spoke perfect English; told me to relax the war was over for me. He looked at my leg and the blood and said I would get medical attention in town. I never did see a doctor or any medical personnel. A truck came by and they put me in the back with a guard and we went into a small town. A piece of the tail section of our plane was in the truck and I lay down on it on the trip to town.

By the time we got to town I had became so stiff and sore it was difficult to get out of the truck as ordered and walk into a room. Once inside I saw Quinten E. Meyer, our ball turret gunner, sitting in a chair. We didn't speak for awhile because a German officer sat behind a desk glaring at us. The officer finally motioned me to a chair in front of the desk and started the questions. All he got from me was name, rank and serial number. He tried the same thing on Meyer and got the same answers. We were put into an army truck and taken to a jail. It was an overnight stop in a cold cell. We were then put on a train and after three or four days were in Frankfort on the Main. We spent a day and a half at the prisoner of war center, then put back on a train, a small boxcar, with a hole in the floor for toilet facilities. We were given very little to eat since being captured. We rode this wonderful accommodation along with about twenty other POWs to a small railroad station called Keifheidi. There we were unloaded and marched a short distance (a mile or two) to the prison camp Stalag-Luft 4. We were again searched and given a Red Cross cardboard suitcase with toilet articles and some new American army clothes.

The stay in Stalag-Luft 4 is well documented in other articles so I won't go into that. The terrible part for most of us started with the now famous death march across Germany. Lager "D" walked out Feb. 6,1945 and this ordeal ended some 500 miles and unrealistic misery later at Bitterfield, Germany on April 26th, 1945. Hunger, sickness, filth, thirst and fatigue took their toll. The 104th Timberwolf Division was the most beautiful bunch of American soldiers I ever saw.

From April 26th to may 5th we were pretty much on our own and ate too much. I got sick and ended up in the 77th field hospital on May 17th at Amiens, France and was taken by ambulance to the 179th General Hospital at Rouen, France. After 27 more days I started for home, and arrived in the states on July 4, 1945 - Bless America. I went to Camp Kilmer, N.J. and called my wife, Althea, who I hadn't seen in over a year and a half. We talked ten dollars worth. I can't describe seeing the Statue of Liberty and hearing my wife's voice. As Lawrence Welk says," Wonderful, Wonderful, Wonderful".

T/Sgt. Shirley W. Hancock<br>U.S. Army Air Force

## DESCRIPTION OF ABBREVIATIONS USED

| | | | |
|---|---|---|---|
| A/CF | Aircraft Factory | HW/B | Highway Bridge |
| Acft | Aircraft | I/A | Industrial Area |
| A/D | Air Drome | IP | Initial Point |
| AM/D | Ammunition Dump | L/D | Locomotive Depot |
| ATT | Attack | MACR | Missing Aircraft Report |
| BOMBS | GP - General Purpose,<br>Frags - Fragmentation<br>RDX - Demolition<br>- Incendiary | MPI | Main point of impact |
| | | M/W | Motor Works |
| | | M/Y | Marshalling Yard |
| BR | Bridge | O/D | Ordnance Depot |
| C/T | Communications Target | O/I | Oil Installations |
| CAVU | Ceiling & Visibility Unlimited | O/R | Oil Refinery |
| C/W | Chemical Works | O/S | Oil Storage |
| E/A | Enemy Aircraft | PFF | Path Finder (By Radar)<br>Also called, "Mickey" ** |
| E/W | Engine Works | | |
| FLAK | S - Scant,,<br>M - Moderate<br>H - Heavy<br>A - Accurate,<br>I - Intense | PT | Primary Target |
| | | RR/B | Railroad Bridge |
| | | RR/S | Railroad Station |
| FW | Fighter Wing | S/D | Submarine Docks |
| G/D | Goods Depot | T/W | Tank Works |
| G/I | Gun Installations | WW | War Weary |
| G/P | Gun Positions | | |
| H/I | Harbor Installations | | |
| Spur | Projection of land northeast of Foggia that would be a spur if Italy were a Boot | | |
| ** | When first developed, radar was called a "Mickey Mouse Operation". The name "Mickey" stuck. | | |

# UNIT HISTORY, 485TH BG (H) 10 MAY 1944 TO 25 APRIL 1945

Following is an account of each of the 187 missions flown by the 485th Bomb Group. At the beginning of each month there is a summary and assessment of the results of the bombing, the losses and the morale of the group. Also listed are the programs that were introduced to keep the morale up.

The first few missions were marshalling yards in Italy and Yugoslavia to gain combat experience with the least casualties. Also practice flights were flown to insure a tight formation during fighter attacks and over the target to get a good bomb pattern. It took guts to keep a tight formation when flying through flak over the target. A direct hit was always anticipated, but since there was no maneuvering room, a plane close by could take a hit causing a collision taking both planes down.

Enemy fighters were seen but not engaged until June 9th when all hell broke loose. Five planes were lost, four to fighters and one to flak. Four of the planes were from the 829th Squadron. After that first encounter, fighter attacks continued on a regular basis through August. Ploesti fell August 30th shutting off about one third of Germany's fuel production. The fighter attacks were few after that. Sometimes they would make one pass through a formation or attack unescorted stragglers.

Ploesti was a tough target and received much publicity, but there were other targets just as lethal. Vienna, with 13 targets had more flak cannons than any city in Europe and then there were Linz, Munich, Blechhammer and Regensburg to name a few. No one was anxious to go to any of those targets.

At full strength, each of the four squadrons had 15 flight crews with 9 or 10 men per crew plus the ground personnel. Each squadron had 15 B-24 Bombers plus a couple of spares. Usually about one half of the planes were sent out on a mission except when a maximum effort was ordered. Each squadron had its own operations office, living quarters (tents), mess hall, etc. Although the squadrons were separated from each other by several hundred yards in case of an air raid, our group was a tight knit community One can understand the shock of losing five crews (48 men) during our first engagement with the Lufwaffe and the anxiety of not knowing if our friends survived. It's hard to explain, but we had to become callused to the losses. Yes, we cared, but we could not let it jeopardize our assignment if we wanted to survive.

In January, after the Battle of the Bulge, everyone knew that Germany was defeated, everyone but Hitler. He ordered the troops to fight to the last man for the Fatherland. In April we bombed the Front Lines in Italy when the final push started up the "Boot". Then we bombed the bridges in northern Italy to cut off the German retreat and our final mission was bombing the marshalling yards in Linz.

## MISSION NO.1 - 10, MAY 1944

Eighteen (18) B-24's bombed KNIN MARSHALLING YARD in Yugoslavia (Primary Target), after the first wave of 18 aircraft turned back because of weather. Thirty-six tons of bombs were dropped from 15,000 feet through 9/10 overcast. Results were generally poor, although several hits were observed on the marshalling yard. Flak was light and inaccurate over the target. No enemy fighters were seen and all aircraft and crews returned without injury.

## MISSION NO.2 - 12, MAY 1944

Thirty-four (34) B-24's attacked the VIA REGGIO MARSHALLING YARD in Italy, with undetermined results, due to inclement weather. 296 500 lb. GP's were dropped. No flak or enemy aircraft were encountered. One crew was forced to ditch on return, killing the Radio Operator and seriously injuring two other crewmembers.
830th Bomb Sqdn Aircraft #148 Crew No. 51
SSgt Hugland (NMI) Hymer - Killed
F/O William S. Lee - Injured
SSgt Harold E. Maxton - Injured
Crew was picked up after spending over an hour in the water.

## MISSION NO. 3 - 13 MAY 1944

Thirty-nine (39) B-24's attacked the MODENA MARSHALLING YARD in Italy and dropped 184 500 lb. GP bombs in the Primary Target. Excellent bomb pattern observed on the yards and factory area. No flak or enemy aircraft were encountered. All aircraft and crews returned to the base safely.

## MISSION NO.4 - 14, MAY 1944

Twenty-nine (29) B-24's attacked the MESTRE MARSHALLING YARD in Italy, dropping 296, 500 lb. GP bombs. The bombing was excellent with bomb strikes plots showing 56% of the bombs falling with 1,000 feet of aiming point. No flak or fighters were encountered and all planes and crews returned safely.

### MISSION NO. 5 - 17 MAY 1944

Thirty-six (36) B-24's attacked the PIOBINO MARSHALLING YARD in Italy, dropping 272, 500 lb. GP bombs with generally poor results. Flak was heavy, moderate and accurate over the target, and no enemy aircraft were encountered. All aircraft and crews returned to base safely.

### MISSION NO. 6 - 18 MAY 1944

Twenty-seven (27) B-24's bombed BRIDGES around NIS, Yugoslavia, with poor results. 132, 500 lb. GP bombs were dropped with a few hits on or near the bridges. Light flak was encountered and no enemy aircraft were seen. All aircraft and crews returned safely.

### MISSION NO. 7 - 19 MAY 1944

Thirty-four (34) B-24's attacked the BOLOGNA MARSHALLING YARD in Italy. 255, 500 lb. GP bombs were dropped from 17,000 feet with unsatisfactory results. Heavy flak was encountered at the target. No enemy aircraft were seen. All aircraft and crews returned without injury.

### MISSION NO. 8 - 22 MAY 1944

Thirty-seven (37) B-24's attacked the VALMONTONE MARSHALLING YARD in Italy. 66, 1,000 lb. bombs were dropped through 9/10 overcast that prevented result observations. Failure of the 2nd wave to find the target decreased the number of bombs dropped. Moderate and accurate flak was encountered and no enemy aircraft were seen. All aircraft and crews returned safely, although 1 officer and 1 enlisted man suffered flak wounds.
831st Bomb Sqdn Aircraft 709, Crew 61
    Lt. Robert V. Tucker - Injured
829th Bomb Sqdn Aircraft 162, Crew 33
    Sgt. Thomas H. Toot – Injured

### MISSION NO. 9 - 23 MAY 1944

Thirty-one (31) B-24's bombed VALMONTONE, Italy for the second consecutive day. Target was obscured by clouds, but 38 tons of bombs were dropped with excellent results. Bomb strike photos showed that 40% of the bombs fell within 1,000 feet of the aiming point. Flak was slight but accurate. No enemy aircraft were encountered and all planes returned to base safely.

## MISSION NO. 10 - 24 MAY 1944

Thirty-six (36) B-24's took off to bomb the WOLLERSDORF AIRDOME at Wiener-Neustadt, Austria, but 28 aircraft failed to release their bombs due to bad weather. Eight aircraft dropped 14 tons of bombs through the overcast with undetermined results. Flak was particularly heavy and accurate over the target area. One aircraft was forced to ditch near the Island of Vis, Yugoslavia. The crew clung to a dinghy while being shelled by the German shore guns for two hours before they were picked up by a British PT boat and taken to the Island of Vis. Three members were injured and the entire crew was rescued. All other aircraft returned to base safely.

829th Bomb Sqdn Aircraft 162, Crew 31

SSgt Harold B. Johnson - Injured
Sgt. Donald A. Ruffer - Injured
Sgt. Stanley M. Pakulla – Injured

## MISSION NO. 11 - 25 MAY 1944

Twenty-two (22) B-24's attacked the AMBERIEU MARSHALLING YARD in France, dropping 36 tons of bombs through the hazy weather. Bombs fell 1500 feet wide of the target. No flak was encountered. Six Me 109's were seen at the target, but did not attack the Group. All aircraft and crews returned safely to base.

## MISSION NO. 12 - 26 MAY 1944

Thirty-five (35) B-24's dropped 73 tons of bombs on the LYON MOUCHE MARSHALLING YARD at Lyon, France. Preceding bombardment by other Groups made it difficult for the 485th to judge their results closely. No flak or enemy fighters were encountered. All aircraft and crews returned without incident to base.

## MISSION NO. 13 - 27 MAY 1944

Thirty-five (35) B-24's dropped 50 tons of bombs on the MARSHALLING YARD, NIMES, FRANCE. Results were obscured by smoke. Sight, inaccurate flak was encountered at the target. No enemy aircraft were seen. All aircraft returned to base safely.

## MISSION NO. 14 - 29 MAY 1944

Thirty-five (35) B-24's bombed the ATZGERSDORF AIRCRAFT FACTORY, Vienna, Austria. Sixty-one (61) tons of bombs were dropped through IAH flak. The target was covered with smoke from preceding bombings. Two of the Group's aircraft were shot down by flak near the target. Twenty (20) parachutes were seen to open and crews are believed to be alive. No attacks were sustained by enemy aircraft and all other aircraft returned to base.

829th Bomb Sqdn Aircraft 797, Crew 36

2nd Lt. Roy E. Daniel
2nd Lt. Edward L. Beecham
2nd Lt. Layton C. Tuggle
2nd Lt. Stevens S. Simmerman
SSgt Mack A. Lundy
Sgt. Otis W. Gammon
Sgt. Frank M. Jasso
SSgt David A. Phillips
Sgt. Daniel P. Haggerty
SSgt E. H. Braziel

831st Bomb Sqdn Aircraft 676, Crew 64

Capt. Joseph P. Landis
2nd Lt. Lloyd E. Proudlove
2nd Lt. Edwin R. Ivey
F/O Julius Q. Tolleson
TSgt Walter K. Gworek
TSgt Wilbur T. Snyder
SSgt Kenneth H. A. Brown
SSgt Robert W. Miller
Sgt. Armond L. Medore
SSgt Arni J. Gudjonsen

## MISSION NO. 15 - 30 MAY 1944

Twenty-two (22) B-24's attacked the AIRCRAFT ASSEMBLY PLANT in NEUNKIRCHEN, AUSTRIA, with excellent results. Thirty-eight (38) tons of bombs were dropped with bomb strike plot showing 57% of the bombs within 1,000 feet of the aiming point. Other bombs fell in a good bomb pattern and factory and warehouses were judged destroyed. The lead pilot, bombardier and navigator were awarded the DFC for this mission. All aircraft and crews returned safely.

## MISSION NO. 16 - 31 MAY 1944

Thirty-one (31) B-24's attacked the PLOESTI REDEVENTA REFINERY, Rumania. The enemy spread the smoke screen across the town and target area so effectively that the crews reported that they couldn't even see one building. Sixty (60) tons of bombs were dropped in the general target area, but no observations could be made regarding the results. Flak was IAH and other Groups were heavily attacked by enemy aircraft. However, all aircraft returned with only one crew member, a gunner, slightly wounded.
831st Bomb Sqdn Aircraft 727, Crew 62.
SSgt Arthur E. Nitsche - Injured

**PLOESTI, ROMANIA.** What else can be said about Ploesti that hasn't already been written? The oil refineries at Ploesti produced 1/3 of the total output of the AXIS fuel. The 485$^{th}$ Bomb Group flew seven (7) missions to Ploesti during the summer of 1944.

## UNIT HISTORY, 485TH BG (H) 1 JUNE 1944 TO 30 JUNE 1944

During this month, the 485th BG (H) attacked a variety of targets in Rumania, Northern Italy, Germany, Hungary, Austria, France, and Yugoslavia. Seventeen (17) missions were flown in all during this period, making a grand total of 33 missions as of 30 June 1944.

1159 tons of bombs were dropped. Results were generally good. In all but three missions, strikes were observed on the target or in the target area.

Fourteen (14) aircraft and their crews failed to return from this month's operations. Of these, one crew bailed out over Yugoslavia. Eight members of this crew fell in the hands of the Partisans almost immediately and were successfully evacuated back to base. Two members are missing.

The Group's score of enemy aircraft encounters were as follows:

    29 Destroyed
    32 Probables
    18 Damaged

Morale of the entire Group continued to be very high. Living conditions were constantly improved. All four Squadrons and the Headquarters Detachment had EM Service Clubs and Officer's Clubs in the process of construction.

Weekly quota was established to send Officers and Enlisted Men to Rest Camps at Villaggio Mancuso, Isle of Capri, and San Spirito.

Several USO shows came to the Group area. An all-soldier show entitled "Somewhere" in Southern Italy" was conceived and presented by Group personnel with great success. The cast included personnel from the 408th Service Squadron of the 323rd Service Group, and from the British Anti-Aircraft Units located on and near Venosa Army Base. This gesture did much to improve the already cordial relations between the three units.

The Group's open-air theatre continued to draw large attendance nightly. Reading of the latest news summary added to the program.

Eight (8) replacement crews were given a thoroughly planned six-day program of indoctrination and orientation. The main subjects covered were: Briefing and Interrogation procedures, Formation Flying, Escape and Evasion, and Aircraft Recognition.

## MISSION NO. 17 - 2 JUNE 1944

At 0530 hours, 34 B-24's took off to bomb the CLUJ MARSHALLING YARD in Rumania (PT). The 1st attack unit was led by Lt. Col. W. E. Arnold, GP CO, and the 2nd attack unit was led by Captain Thomas D. Obrien, 829th CO. One aircraft returned before assembly due to landing gear malfunction.

The group rendezvoued with the fighter escort, P-51's of the 306th FW, at 0901 hours. The fighter remained with the bomber formation through the target area. One aircraft returned prior to bombing due to turbo trouble, at 0954 hours. No enemy aircraft or flak was encountered on the route to the target.

Thirty-two (32) aircraft were over the target at 0923 hours and dropped 63.5 tons of 500 lb. GP bombs. Some bombs, dropped on the first run, were observed to hit the east end of the marshalling yard in the city of Cluj. The bombs dropped on the second run were observed to fall in the target area which was, however, obscured by smoke and results could not be seen. Crews believed that bombs of the second run hit in the target area.

No aircraft were damaged and there were no casualties.

The weather over the target was mostly clear, with a few scattered clouds. There was a slight to moderate haze.

## MISSION NO. 18 - 4 JUNE 1944

At 0620 hours, 31 B-24's (36 scheduled) took off to bomb the TURIN MARSHALLING YARD in Northern Italy (PT). The 1st attack unit was led by Lt. Col. William L. Herblin, Deputy GP CO, and the 2nd attack unit was led by Major Edward H. Nett, 828th CO. Assembly was over Altamura at 9,000 feet at 0700 hours. Fighter escort was first seen at 0837 hours at 15,000 feet. P-38's, P-47's and P-51's provided penetration to the target and withdrawal cover.

At 1026 hours, 58 tons of 500 lb. GP bombs were dropped. Target was well covered by a good bomb pattern. Several sticks of bombs were seen to fall across the complete width of the marshalling yard. Some bombs, which fell short over the marshalling yard, were seen to hit car sheds and factory buildings.

There was 6/10 to 8/10 cumulus at 6,000 to 8,000 feet in the target area, but a hole in the undercast opened up and bombers approached the target allowing good visibility of the aiming point.

There was scant, inaccurate to scant, accurate, heavy flak over the primary target. One aircraft returned with two small holes and another with one small hole caused by flak. There was no enemy aircraft encountered or seen over the target or on return.

Return was made without incident. Thirty-one (31) aircraft landed safely at 1339 hours.

## MISSION NO. 19 - 5 JUNE 1944

At 1010 hours, 35 B-24's scheduled, took off to bomb the FORLI MARSHALLING YARD (PT) in northern Italy. Thirty-three (33) aircraft assembled at Altamura at 8,000 feet at 1052 hours. The 1st attack unit was led by Major Griffin, 830th CO and the 2nd attack unit was led by Captain Sjodin, 831st CO. No fighter escort to or from the target. No enemy aircraft encountered or no flak encountered. Sixty-four (64) tons of 500 lb. GP bombs were dropped from 17,000 feet. The first attack unit dropped their bombs approximately one to two miles over the target in an agricultural section. The second attack unit successfully attacked the primary target, some hits being observed short of the marshalling yard; approximately 25 hits on the tracks were observed, and some on the buildings beyond the marshalling yard.

The weather was CAVU.

Return accomplished normally and all 33 aircraft landed without incident at 1500 hours.

## MISSION NO. 20 - 6 JUNE 1944

Thirty-five (35) B-24's took off to bomb the DACIA ROMANA OIL REFINERY (PT) at Ploesti, Rumania. The 1st attack was led by Major Robert E. Smith, GP Air Inspector and the 2nd attack unit was led by Captain John B. Stoddart, 830th Operations Officer. Fighter escort provided target, penetration and withdrawal cover back to the Yugoslavian coast. Due to a malfunction of the bomb release mechanism in the lead aircraft, bombs of this aircraft were salvoed just past the IP. The lead aircraft maintained the course from the IP, after the malfunction of bomb release mechanism, over the PT in order that the other aircraft could bomb the target. Due to a very effective smoke screen, making identification of the target was extremely difficult, most of the aircraft of the 1st attack unit and some of the 2nd unit dropped bombs on a line from the IP to the target, with some of the last bombs being dropped in the target area. Fifty-two (52) tone of 500 lb. GP bombs were dropped. Results of the bombing were not observed, due to almost complete smoke screening of the target area.

About 15 enemy aircraft were seen in the target area and immediately beyond the target. These were 8 Me 109's, 6 FW 190's and a twin-engine single tail fighter, probably a Me 210/410. None of these aircraft attacked the formation, but several single engine fighters were seen to attack another formation with unobserved results. The flak, just before and over the target, was moderate to intense, mostly inaccurate, heavy. Several very minor incidents of flak damage were sustained by the Group and one minor casualty, a slight mouth wound, was incurred by flak.

The weather enroute and over the target and return was slightly hazy, 3/10 overcast and 5/10 undercast.

Return was made without incident by 31 aircraft landing at 1300 hours. One aircraft landed at Bari, to have to injured man (mentioned above) attended to and then proceed to base. No aircraft were lost and none are missing.

## MISSION NO. 21 - 7 JUNE 1944

At 0625 hours, 34 B-24's (36 scheduled) took off to bomb the LA SPEZIA HARBOR INSTALLATIONS in northern Italy. The 1st attack unit was led by Lt. Col. Arnold, GP CO and the 2nd attack unit was led by Captain Francis P. Dalton, 829th Flt Cmdr. Fighter escort - P-38's - provided target cover and remained with the formation until Corsica was reached at 1049 hours on the return route. One aircraft returned early. Scant to moderate accurate heavy flak was encountered over Leghorn. One crew reported seeing red bursts of flak in the Leghorn area. There were no encounters with enemy fighters. One Me 109 was seen over Leghorn at a distance.

Finding the target completely overcast, 33 aircraft were over the alternate target of Leghorn at 1029 hours, dropping 80.5 tons of 1,000 lb. GP bombs. Results were good. There were broken cumulus at 8,000 feet enroute to the target. At landfall a complete undercast was encountered, obscuring the IP and the PT at Leghorn.

At 1028 hours, 5 fighters were seen taking off from an airdrome near Pisa, and two twin-engine aircraft were observed on the ground. Several smoke pots were burning over the waterfront at Leghorn. These pots did not impair the visibility for the bombing.

Return was made without incident, 33 aircraft landed at 1300 hours.

## MISSION NO. 22 - 9 JUNE 1944

Mission 22 on June 9th was a bad day for the Group on a mission to bomb an AIRCRAFT FACTORY, MUNICH, GERMANY.

Between enemy fighters and flak, the Group lost 5 aircraft from the 829th squadron and one from the 830th.

Legend:  MACR  Missing Aircraft Report
+ Downed by Enemy Fighters
++ Downed by Flak

+ Crew 22 – Plane Serial #41-28782H

Crew Chief – M/Sgt. Earl L. Bundy
MACR – 6042 – shot down near INNSBRUCK
P      1st Lt. Hugh B. White – POW
CP     2nd Lt. Charles Duecker – POW
N      2nd Lt. Charles A. Field – POW
B      2nd Lt. John C. Norris – POW
E      TSgt. James L. Gillet – POW
RO     Cpl. Ben Thompson (NMI) – POW
AG     SSgt. John H. Hawk – POW
G      Sgt. Benjamin L. Thigpen – POW
G      Gilbert R. Lish – POW
G      ???


++ Crew 32 – Plane Serial # 41-29497H
Crew Chief - SSgt. Christopher O'Keefe
MACR – 5643 – Shot down over target near VEIDEN

P      F/O J. A. Laitwaitis
CP     F/O Elmer D. Kohler
N      2nd Lt. Morris Burney (NMI)
B      2nd Lt. Marion E. Shelor
E      Sgt. Otis H. Vinson

| | |
|---|---|
| RO | Sgt. Jack D. Mizrahi |
| AG | Sgt. Edgar A. Pierce |
| G | Sgt. Edward Walz (NMI) |
| G | Sgt. Simon J. Ventamiglia |
| G | ??? |

+ Crew 33 – Plane Serial # 42-52742H
Crew Chief - Unknown??
MACR – 6006 – Shot down near STRAUBING

| | |
|---|---|
| P | F/O John D. Bond |
| CP | F/O Albert B. Roman |
| N | 2nd Lt. Joseph Duffield |
| B | 2nd Lt. Edward C. Sawyer |
| E | SSgt. Albert W. Knott |
| RO | Sgt. Clarence R. Nippes |
| AG | S/Sgt. William D. Whorton |
| G | Cpl. Thomas H. Toot |
| G | S/Sgt. Elliot B. Altshuler |
| G | ??? |

+ Crew 35 – Plane Serial # 42-78141G – Plane Name MISSTIT
Crew Chief - MSgt. William G. Tanner
MACR - 6065 – Shot down near SIEGENSBURG

| | |
|---|---|
| P | 2nd Lt. Joseph W. Cathecart – POW |
| CP | 2nd Lt. Slayton McGhee – POW |
| N | 2nd Lt. Arthur Carlson – POW |
| B | 2nd Lt. Donald R. Rohen – POW |
| TTG | Sgt. Allen G. McBride – KIA (killed in action) |
| BTG | SSgt. Dewey Holcomb – POW |
| NTG | Sgt. Leon Best – POW |
| RO | Sgt. Mervin H. Lindsay – POW |
| E | Sgt. Roy Mehrkens – POW |
| UTG | Sgt. Irvin R. Wolf – POW |

+ Crew 69 – Plane Serial # 42-52719H
Crew Chief - MSgt. C.L.Olney
MACR - 5482 – Shot down near VILLANOVA
This was a strange situation - It was the 1st 830th plane lost in combat. The pilot & crew assigned for the mission was from the 831st squadron

| | |
|---|---|
| P | 2nd Lt. James C. McNulty |
| CP | 2nd Lt. Eugene E. Maylath |

| | |
|---|---|
| N | F/O Ormiston D. Brown |
| B | 2nd Lt. Edward P. Lubanovich |
| E | SSgt. Alfred P. Bertelli |
| RO | SSgt. Lawrence P. Griggs |
| AG | Sgt. Robert F. Irmen |
| G | Sgt. Eugene J. Brittin |
| G | Cpl. Murray C. Sheridan |

## MISSION NO. 23 - 10 JUNE 1944

At 0615 hours, 32 B-24's took off to bomb the TRIESTE MARSHALLING YARD (PT) in northern Italy. The 1st attack unit was led by Major Edward H. Nett, 828th CO and the 2nd attack unit was led by 1st Lt. John M. Jones, 831st "B" Flt Cmdr. Fighter escort consisted of P-38's, P-47's, and P-51's. There were no early returns.

Fifteen (15) to twenty (20) enemy aircraft were seen between Grado and Monte De Cap: 4 FW 190's, 4 JU 88's, 1 Me 210 and the remainder Me 109's. Six enemy aircraft attacked the formation over the target area.

Finding the target partly obscured by the clouds, only 12 aircraft dropped 30 tons 500 lb. GP bombs. A right turn was made and a second run was made on the target at 22,000 feet. Nineteen aircraft dropped 47.5 tons of 500 lb. GP bombs. One aircraft failed to drop because of malfunction. Hits were observed in the marshalling yard, some on the oil storage facilities, which caused flames to leap to a height of approximately 1500 feet. A few bombs fell over the target, falling in a residential district. Several bombs were observed to have fallen into the water, short of the target, and some possibly hitting a pier.

SIH flak was encountered near the IP. At the target, SIH flak was again encountered. Some red bursts were observed, at which time the flak ceased and the enemy aircraft attacked. These red bursts were apparently used as signals. About 6 enemy aircraft were encountered over the target area, being equally divided as two types, FW 190's and Me 109's. These aircraft attacked individually and in pairs from 8 and 6 o'clock, high and level. Attacks were sporadic from approximately 0945 to 0959 hours. Pilots appeared to be experienced and were aggressive in that one attack was pressed within 200 yards

Bombers claimed 1 Me 109 probably destroyed and 1 Me 109 damaged. Enemy aircraft claimed as damaged, attacked from 6 o'clock low. Bomber opened fire at 700 yards and fired about 100 rounds into the enemy aircraft before it broke away at 200 yards with smoke pouring from it. It was last seen losing altitude, but apparently under control. The enemy aircraft claimed as probably damaged, came in from 6 o'clock low. Bomber opened fire at 1,000 yards. Two other enemy aircraft followed this aircraft in. Bomber fired 150 rounds at one of these. Tracers were seen to go all around the cockpit with the enemy aircraft flying through a virtual wall of fire. The top gunner verifies the first of these encounters, having seen the enemy aircraft stop suddenly and go into a spin.

The weather enroute was hazy with very little cloud cover. Over the target at 0930 hours, a large cloudbank at 15,000 feet obscured the target area. At 1001 hours, when the 2nd run was made, the cloud layer had moved and the target was partially visible.

Return was made without incident, landing at 1233 hours. There were no stragglers, no late arrivals, no losses and no casualties. One aircraft reported one small hole from flak. No bombers were damaged seriously by either flak or from enemy aircraft fire.

## MISSION NO. 24 - 11 JUNE 1944

At 0600 hours, 35 B-24's (37 scheduled) took off to bomb the SMEDERVO OIL REFINERY and TRAIN FERRY (PT) in Yugoslavia. The 1st attack unit was led by Major Robert E. Smith, GP Air Inspector and the 2nd attack unit was led by Major Richard E. Griffin, 830th CO. Lt. Col. Tomhave, Major Smith and 1st Lt. Conway, 55th BW Officers, flew with 1st Lt. Jones, 2nd Lt. Tuttle and 2nd Lt. Skelton's crews respectively. There was no fighter escort or early returns. No enemy aircraft were encountered or seen.

Over the target at 0905 hours, 34 bombers dropped 84.5 tons of 500 lb. GP bombs. One aircraft did not drop because of malfunction. The target was well covered by a concentration on the main cracking plants. Last aircraft over the target reported large fires in the target area, which was well covered by smoke. No flak was encountered over the target or at any point enroute or on the return trip.

Weather was good.

Return was made with incident, landing at 1145 hours.

## MISSION NO. 25 - 13 JUNE 1944

At 0603 hours, 38 B-24's (39 scheduled) took off to bomb the MILBERTSHOFEN ORDNANCE DEPOT (PT) at Munich, Germany. The 1st attack unit was led by Lt. Col. Arnold, GP CO and the 2nd attack unit was led by Major Richard V. Griffin, 830th CO. Two groups of P-38's provided fighter escort over the target and upon withdrawal, departing at 1150 hours, well over the Adriatic Sea. Aircraft's 601, 503, 127, 532, 701, 694, 730, 699, 720 and 721 returned early. Early Return Board met at 1500 hours and discussed the circumstances in each case; eight were legitimate and two are still in question.

Eighteen (18) enemy aircraft, 15 Me 109's and 3 FW 190's were seen in the area of Lake Chiem to the IP. Moderate, heavy flak was observed. The flak was moderate to intense over the entire Munich area.

Twenty-eight (28) aircraft were over the target at 1029 hours and 24 aircraft dropped 48 tons of 500 lb. GP bombs. Two aircraft jettisoned 4 tons of bombs approximately 2 minutes before the IP in a wooded area, due to malfunction.

Observations of bomb strikes were not made due to the target being well covered by a smoke screen. The last aircraft over the target reported large clouds of rolling smoke, as from burning buildings.

About 11 enemy aircraft attacked the formation between the IP and the target. The attacks were not aggressive, being made singly and in threes, from 3, 6, and 9 o'clock high and 6 o'clock low. Bombers claimed 1 Me 109 damaged.

One bomber, 118, was observed to have been hit by flak over the target at 1029 hours. Eight (8) parachutes were seen to open and descend in the vicinity of the Munich area. It is believed that the other two crew members made delayed jumps.

Another bomber, 709, was last seen at 1033 hours near Anzing when it was straggling and being attacked by enemy aircraft. This aircraft is reported as missing since no one saw it go down.

An airdrome at Munich had many single engine fighters on the ground at target time. Two Me 110's and 2 B-17's were observed on a field just east of the target near Anzing.

Markings on enemy aircraft were - yellow nose, upper fuselage gray, underside cream colored and white wing tips.

The weather was hazy over the water with scattered cumulus over the Alps.

Return was made without incident, 26 aircraft landed at 1327 hours and 2 aircraft

## MISSION NO. 26 - 14 JUNE 1944

At 0750 hours, 39 B-24's, scheduled, took off to bomb the PETFURDO OIL PLANT (PT) in Hungary. The 1st attack unit was led by Lt. Col. Wm. L. Herblin, Deputy GP CO and the 2nd attack was led by Major Edward H. Nett, 828th CO. The fighter escort consisted of P-38's. Aircraft 727 returned early.

Thirty-eight (38) aircraft were over the target at 1103 hours of which 35 aircraft dropped 69.37 tons of 250 lb. GP bombs. Target was well covered by smoke from bombing by other Groups ahead of the formation. Results of the bombing were undetermined, although some hits were observed on oil storage tanks slightly northwest of the target.

Over the target about 5 enemy aircraft were seen. None of these attacked the bomber formation. Several crews reported two enemy aircraft shot down - 1 enemy aircraft exploding and another going down out of control after the pilot had bailed out at 1102 hours. Another crew reported seeing an enemy aircraft shot down at 1109 hours. SAH flak was encountered over the target. Four bombers received small holes from flak. One bomber, 504, had it's no. 4 engine knocked out by a bomb dropped from an aircraft ahead, exploding in the air within 400 feet of the bomber. No aircraft lost, none missing, none down at friendly fields and no casualties. The weather was CAVU enroute and over the target. Return was made without incident, 38 aircraft landing at 1347 hours.

## MISSION NO. 27 - 16 JUNE 1944

At 0608 hours, 39 B-24's scheduled, took off to bomb the LOBAU OIL REFINERY (PT) at VIENNA, AUSTRIA. The 1st attack unit was led by Major Robert E. Smith and the 2nd attack unit was led by Captain Daniel L. Sjodin, 831st CO. P-38's, P-47's, and P-51's provided cover over the target and on return, to a point just beyond the Lake Balton area, although they had become scattered due to enemy attacks.

Five aircraft returned early: 703, 729, 769, and 132.

Between 30 and 40 single-engine enemy aircraft, mostly Me 109's and 14 twin-engine fighters were seen in the target area. These enemy aircraft attacked both units of the formation. They pressed their attacks vigorously, mostly against the second unit. The 1st unit had 11 enemy aircraft start in from 11 o'clock, but they did not attack as they were intercepted by P-38's. There were approximately 38 passes made by enemy aircraft on both units, mostly from 6 o'clock, slightly low. None of our aircraft were shot down by enemy aircraft or by flak, but the following observations were made of losses by other Groups participating: One B-24 blew up over the target; another B-24 exploded from which one parachute was observed; another B-24 was seen going down with no one bailing out; and a P-38 was observed to be going down.

One Me 109 crashed. Eleven (11) P-38's were seen to shoot down a Me 210. A P-51 shot down a Me 109, which blew up in the air. Five Me 109's and One Ju 88 were seen going down at Lake Balton. Two (2) or our aircraft landed at the Isle of Vis after their ships had been damaged by enemy flak incurred over the target. Gunners of our Group claimed: 5 Me 109's destroyed; 2 probables, 4 damaged. Markings of enemy aircraft were observed as follows: Me 109's - silver with black crosses, red, white, and green triangle on Me 110 and Me 109's with red and yellow wing tips. Flak was IAH over the target.

Thirty-four (34) aircraft were over the target at 1033 hours and dropped 66.37 tons of 250 lb. GP bombs from 23,000 feet. The target was covered by a dense smoke. The Bombardier had to synchronize on smoke, which was drifting. The second unit encountered the same trouble. Bombs were seen to strike into the smoke. The weather was CAVU.

Thirty-one (31) aircraft landed at the base safely at 1320 hours. One aircraft landed at Bari, Italy, leaving one wounded crewmember who had sustained a wound in the neck, at the 26th General Hospital, for treatment. This was the only casualty on the mission. Aircraft returned to base at 1320 hours. Two aircraft landed at the Isle of Vis due to aircraft being badly damaged by flak. One aircraft was hit by flak, losing fuel in no.1 and no. 4. No. 3 engine gave trouble on way up and set up a terrific vibration; power setting had to be cut to 30 inches in order to maintain this engine. Aircraft landed at Vis at 1245 hours and the crew was returned to Bari where they were picked up by an aircraft from this Group and returned home airdrome at 2030 hours. They reported that the 2nd crew on Vis was injured.

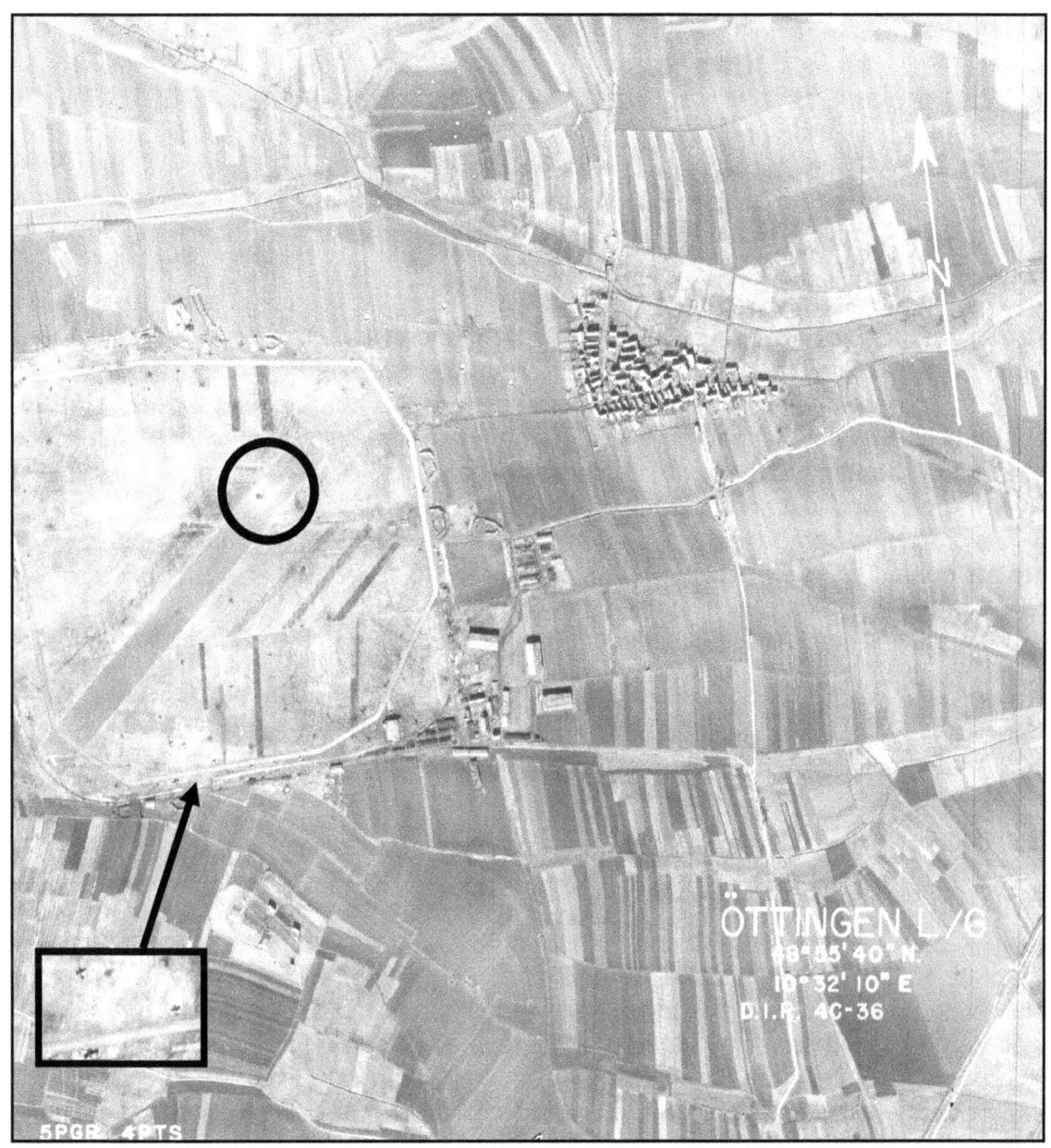

## OTTINGEN L/G GERMAN FIGHTER PLANE LANDING STRIP

This fighter strip was located between Nurnberg and Augsburg. There were fighter strips like this one scattered throughout Germany. When they heard the roar of the bombers approaching, they scrambled to meet them. A taxi strip surrounds the runway and the planes are widely dispersed around it. There are about fifty (50) fighter planes in the photo. A bomb crater can be seen on the runway.

## MISSION NO. 28 - 22 JUNE 194

At 0752 hours, 39 B-24's, scheduled, and took off to bomb the CASTEL-MEGGIORE MARSHALLING YARD in northern Italy, (PT). The 1st attack unit was led by Major Edward H. Nett, 828th CO and the 2nd attack unit was led by 1st Lt. Francis P. Tunstall, 830th Flt. Cmdr. P-51's provided cover for the formation to the target. P-38's were observed over the target and furnished cover upon withdrawal to a point off the coast opposite Ancona, Italy. There were no early returns.

Fifteen (15) enemy aircraft, apparently ME109's were seen between Formignana and the target, but they did not approach the formation. Moderate - heavy flak was observed at Bologna. Scant-heavy flak was observed at Pradurg e Sasso, Ferrara and Vergato.

Thirty-nine (39) aircraft were over the target at 1114 hours and dropped 78 tons of 500 lb. GP bombs. The 1st unit got some hits on the briefed mean point of impact with some bombs falling short and to the left of the target. The bombs of the 2nd unit fell to the left of the marshalling yard and just short of main railroad.

After rallying from the target, the formation zigzagged, dodging flak bursts to Rimini, where the briefed course was followed back to the base.

Weather was CAVU with scattered cumulus.

Return was made without incident, 39 aircraft landing at 1407 hours. No aircraft lost and no casualties.

## MISSION NO. 29 - 23 JUNE 1944

At 0611 hours, 38 B-24's (39 scheduled_ took off to bomb the IRGIU OIL INSTALLATIONS AND STORAGE (PT) at Guirgiu, Rumania. The 1st attack unit was led by Lt. Col. Walter E. Arnold, GP CO and the 2nd attack unit was led by Captain Thomas D. O'Brien, 829th CO. Six P-51's and 12 P-38's provided fighter escort. One aircraft lost the formation and returned early due to weather and joined the 304th Wing, which turned back. The aircraft landed at base at 0941 hours. Five other aircraft returned early due to malfunctions.

Moderate-accurate-heavy flak was encountered over the target for approximately 3 minutes. Some railroad flak was observed south of the target. One (1) aircraft was lost to flak over the target area.

Thirty-two (32) aircraft were over the target at 0945 hours and dropped 64 tons of 1,000 lb. GP bombs. Bomb strikes were observed in the target area; some dropped short and some were seen to go over the target. No heavy smoke was observed.

About 15 enemy aircraft were seen between the IP and the target and 30 miles west of Sofia, on return, of which 8 were Me 109's, 4 FW 190's and 3 believed to be JU 87's as they had fixed landing gear. At 1101 hours, one Me 109 made a pass at the 2nd attack unit - the attack came in at 3 o'clock low and broke away at 8 o'clock, but was unaggressive. Gunners fired with no results. Color noted on enemy aircraft: Me 109 - Silver gray in color - some with yellow noses - one with red nose and black and white fuselage with white swastika.

One aircraft 495, was lost by a direct flak hit - bomb bay was in flames and flames were also coming out of the upper gun turret. Aircraft went into a 45 degree dive out of the formation, then pulled out and was last seen under control. Eight (8) men were seen to come out of the aircraft, 6 parachutes definitely opening. Red flares were fired out of the aircraft, on the way down. Aircraft was at approximately 21,800 feet when hit.

Aircraft 706 is missing. When last seen, it dropped out of formation and turned bask 45 miles west of the target. Radio report from aircraft 706 received - two engines out, heading 240 degrees.

Target area was 2 - 3/10 cumulus with tops at 6,000 feet. Route same as out with cumulus building up to 18,000 feet over all mountains.

Return was made without incident, 32 aircraft landing at 1311 hours.

## MISSION NO. 30 - 25 JUNE 1944

At 0545 hours, 35 B-24's (39 scheduled) took off to bomb the SETE OIL REFINERY (PT) at Sete, France. The 1st attack unit was led by major Richard V. Griffin, 830th CO and the 2nd attack unit was led by Captain Van E. Neal, 828th "C" Flt Cmdr. Twenty-five (25) P-28's provided fighter escort.

Flak at the target was SIH - some railway flak was observed...Formation was exposed to flak for approximately 2 minutes.

Eighteen (18) aircraft of the 1st unit were over the target at 1024 hours and dropped 36 tons of 500 lb. GP bombs. Target was hit - strikes were seen in the target area. Black smoke and flames billowed up to about 1,000 feet. Seventeen (17) aircraft of the 2nd unit were over the target (CHEMICAL WORKS AND OIL STORAGE AT SETE, SOUTHWEST OF PT) at 1025 hours and dropped 34 tons of 500 lb. GP bombs. Second unit did not bomb the PT because they identified it too late to do any satisfactory bombing; therefore, continued on course and bombed the alternate target. Strikes were observed in the target area - a tanker was hit in the lagoon. Smoke was seen in the target area of the Chemical Works and Oil Storage facilities.

The weather over the base at take off was 8 to 9/10 strato-cumulus at 9,500 feet. Over the target it was clear. On return, cumulus over the Appenines built up to 14,000 feet with 7 to 10/10 over the base.

There were no claims, losses or casualties. No enemy aircraft were encountered.

Return was without incident, 35 aircraft landing at 1430 hours.

## MISSION NO. 31 - 26 JUNE 1944

At 0535 hours, 38 B-24's scheduled, took off to bomb the FLORISDORF OIL REFINERY (PT) at Vienna, Austria. The 1st attack unit was led by Captain Daniel D. Sjodin, 831st CO and the 2nd attack unit was led by Captain John C. Sandall, 829th Flt Cmdr. Six groups of P-38's of the 306 Wing provided fighter escort at 0825 hours. Fighters provided penetration, target and withdrawal cover as far as the Yugoslavian coast. As the bomber formation crossed the coast on return, several P-38's were seen covering straggling bombers. Four aircraft return early.

One aircraft, which is missing, was seen to turn back with one engine, no. 4, smoking slightly, at 0910 hours, just north of Lake Balaton. Four unidentified silver colored fighters were seen following the aircraft, but it was impossible to determine whether these were friendly or enemy aircraft.

About 35 to 40 enemy aircraft: JU 88's, Me 110's, Me 210's, Me 109's and FW 190's were encountered between the IP and the target. Attacks were aggressive and continued over the target. Enemy aircraft attacked from 11 o'clock low and level with the formation, passing right through the formation and firing rockets, incendiary rockets, as well as cannon and machine guns. Several aircraft received minor to moderately severe damage from machine gun and cannon fire. Crews claimed 8 enemy aircraft destroyed - 4 Me 109's, 2 Me 110's, 1 Me 210/410, and 1 FW 190; and two probables - Me 210/410's. Marking of enemy aircraft: Me 110's and Me 210's, yellow nose, Me 210's with blue and yellow fuselage and dark blue or black cross on the side of the fuselage. Me 210's dark bridle camouflage, Me 110's with bluish gray spirals on prop, Me 109's with red spinners, Me 109's with silver with orange tails, Me109's with black fuselage and red tails. Number of different types were approximately: 8 JU 88's, 15 to 20 Me 110's and Me 210/410's and approximately 15 Me 109's and FW 190's. Although the first attack was mainly from the front, succeeding passes were made at the formation from 5 to 6 o'clock.

Thirty-three (33) aircraft were over the target at 0945 hours and 31 aircraft dropped 60.5 tons of 100 lb. Incendiaries from 22,800 feet. Target area was well covered and hits were seen in the marshalling yard.

IAH flak was encountered in the target area. One aircraft of our formation believed to be hit by flak, went down in the target area. Seven chutes were definitely seen and 3 others possibly. Most of the aircraft in the formation received minor to moderate incidents of flak damage.

One aircraft is lost, probably to flak over the target, and one aircraft is missing. One aircraft is reported to have landed at the Isle of Vis. Co-pilot of one aircraft was injured over the target area. The injury was moderate and is believed to have been caused by flak.

Weather enroute and over the target was CAVU.

Thirty (30) aircraft landed at 1255 hours. One aircraft landed at Bari, Italy at 1240 hours with wounded crewmember and returned to base at 1410 hours.

## MISSION NO. 32 - 28 JUNE 1944

At 0548 hours, 38 B-24's (39 scheduled) took off to bomb the TITAN OIL REFINERY AND STORAGE at BUCHAREST, RUMANIA, (PT). Major Robert E. Smith, GP Ops Officer, led the 1st attack unit and Captain Ralph E. Monroe, 828th Ops Officer, led the 2nd attack unit. Twenty (20) P-38's and some P-51's rendezvous with the formation at 0807 hours and left the formation at 0945 hours. These fighters were not seen again until 1023 hours (20 P-38's) and again left the formation at the Yugoslav coast on the return flight. Two aircraft returned early.

Thirty-three aircraft were over the target at 0952 hours and dropped 82 tons of 500 lb. GP bombs. The 1st unit, consisting of 16 aircraft, dropped 39.5 tons of bombs on the first alternate target - a factory in the city of Bucharest was used as the 1st unit's aiming point.

Many strikes were seen in the target area and fires started. The primary target was obscured by clouds. The 2nd unit, consisting of 17 aircraft, dropped 42.5 tons of bombs on the primary target. Strikes were seen in the target area and some fires started.

About 45 to 50 enemy aircraft were seen between the IP and the target, of which 10 were FW 190's, 30 Me 109's, 1 JU 88 and 6 Me 110's and JU 87's. About 32 Me 109's and 3 FW 190's attacked the formation. Attacks were continuous from the IP to the target, and for 30 minutes. Attacks were aggressive, coming in from all around the clock, high, level and low. Enemy aircraft came in from 2 to 6 abreast, and 2 to 4 in line. The pilots appeared to be quite experienced. Bombers claimed 12 Me 109's destroyed, 15 Me 109 probables and 1 FW 190 damaged. Markings: Me 109's - dark gray in color, several black and silver with swastika on fuselage and tail, some with yellow ring around the fuselage with yellow tail markings and yellow spinners.

Flak at the target was IAH. Aircraft of the formation which were holed by flak: 7 - slight, and 5 - severe. Six aircraft of the formation were holed by fighters: 5 - severe and 1 - Minor. Three of our bombers were lost due to enemy aircraft. Aircraft 147 was seen to go down shortly after passing over the target. Four chutes were seen to open. Aircraft 701 was seen to go down and 6 chutes were seen to open. Aircraft 122 was seen to go down shortly after passing over the target. Two men jumped out and one chute was definitely seen to open.

Aircraft 122, which was flying no. 6 position in the low box of the second unit, was attacked by 4 Me 109's, at which time no. 3 engine received a direct hit and almost blew out of the nacelle, after which the aircraft dropped out of the formation. At that moment the aircraft was attacked again by the 4 fighters and the aircraft was seen to go in a vertical dive. Two men were seen to jump out and one chute was seen to open, definitely. The aircraft then caught fire and hit the ground. Parts were seen flying off the aircraft while it was in the dive. The time was 1004 hours.

Aircraft 701, which was flying no. 7 position in the lead box of the 2nd unit, was hit by what appeared to be 20-mm shells, in the waist section. One 20-mm shell was seen to hit in the fuselage near the co-pilot's seat. Aircraft was seen to drop out of formation with 3 Me 109's following. The aircraft was seen to hit the ground and explode at approximately 1003 hours.

Aircraft 147, which was flying no. 6 position in the low box of the lead unit, was attacked by 3 Me 109's, then was seen to straggle out of formation under control. No critical damage was visible. Then, 5 to 7 enemy aircraft attacked the B-24. It caught fire, went into a spin and crashed. Before the aircraft crashed, 4 men were seen to bail out and 4 chutes were seen to open. The aircraft crashed at approximately 1000 hours.

Eleven men were wounded in aircraft 534 and 703. Casualties on aircraft 116 have not been reported at this writing.

2nd Lt. Volney W. Wiggins, was the pilot of aircraft 534, which was attacked by enemy aircraft after the target. Five crewmembers were injured and the aircraft had approximately 400 20 mm and shrapnel holes. Co-pilot, Mathew W. Hall, 2nd Lt., distinguished himself by continuing to assist the pilot in keeping the badly damaged aircraft in flight, although he was seriously wounded. The uninjured crewmembers, other than the pilot, 2nd Lt. Kenneth S. Leasure, T.Sgt William B. Shimer, Cpl. Edward L. Hartupee, performed an excellent job in administering first aid to the injured crewmembers. Their efforts, knowledge and ability in administer first aid saved three of the seriously wounded crewmembers from bleeding to death.

1st Lt. Young B. Barber, flying aircraft 414, dropped out of formation to protect aircraft 534 from enemy aircraft. He escorted the damaged aircraft from 1015 hours until it reached Bari at 1300 hours.

The weather enroute was broken cumulus clouds.

Twenty-nine (29) aircraft landed without incident at 1330 hours. Aircraft 534, 703 and 718 landed at Bari, Italy. Aircraft 116 landed at Tortorella, Italy. Aircraft 718 returned to home base, the others remained at the friendly airfields as they were severely damaged.

## MISSION NO. 33 - 30 JUNE 1944

Thirty-six (36) B-24's, scheduled, took off to bomb the BLECHHAMMER OIL REFINERY (PT) in southeastern Germany, but bombed a target of opportunity - SPLIT MARSHALLING YARD and HARBOR INSTALLATIONS, on the coast of Yugoslavia. Major E.H. Nett led the 1st attack unit and Captain John C. Sandall, 829th Flt Cmdr, led the 2nd attack unit. There was no fighter escort. Aircraft 812, 498 and 299 returned early.

Sixteen (16) aircraft of the 1st unit were over the target of opportunity at 1120 hours and dropped 39.74 tons of 500 lb. GP bombs from 20,000 feet. Strikes were observed in the marshalling yard and harbor installations at Split.

Seventeen (17) aircraft of the 2nd unit returned bombs to base.

No enemy aircraft were seen. There were no losses.

Weather was clear along the entire route north as far as Lake Balaton, and the rest near overcast at 11,000 feet.

Return was made without incident, 33 aircraft landing at 1230 hours.

# Letter Home

The following letter was written on August 12, 1944. Although it is unsigned, I am assuming it was written by a friend of our former Historian and War Buddy, Carl Gigowski who died of Leukemia in 1993. The letter covers a multitude of stories; I feel you will find very interesting from a Historical and Educational viewpoint even though we do not know the people he relates to. I found this letter in the many memorabilia sent to me in preparation to write The History Of The 485th. Sammy Schneider

Dear Hansells and Hansens,

I think I will make this a joint letter as I think it will be fairly long and I don't think you folks would mind getting it that way. I got this idea from Aaron as he writes to Four of us in duplicate.

I really feel ashamed for not writing for so long, but it seems like all the writing that I have been doing for the past month has been to Babin and my folks. If I have an excuse, I believe it is that I haven't been feeling too good. I have a lot of trouble with my back and get very tired if I sit and write too long. I think it might be some form of rheumatism, as I am so stiff and sore. My neck and shoulders are so stiff all the time. I don't do any lifting as when I bend over, I can hardly get up. We have been having plenty of rain the past two weeks: and I believe that may have something to do with it. Yesterday we had quite a down pour from three in the afternoon till eight last night. Our tent does a pretty good job of leaking and I have quite a job trying to find a dry spot for my cot. My blankets got pretty damp last night. Well how does the war news sound to you folks. Although the Russians have slowed down, I believe there is a method in their madness. After making gains like they did, they have to slow down to bring up supplies, etc. I look for them to steam roll again pretty soon. In Italy, I believe they are trying to hold what Germans they have there so they can't fight on other fronts. I really don't believe the Allies would ever get into France by way of Italy on account of the rugged country.

Maybe the French Army will hit Southern France. Our news this afternoon although unofficial say the Americans are within 50 miles of Paris. Hope it all continues to go well, as I believe it will. I don't know what the setup will be, after Germany surrenders. There will be no use that I can see of us staying here. There is quite a lot of thought here that we may go back to the States to train in a B-29 Squadron but that is just a rumor and they aren't worth a dime a dozen. If the whole thing would be over before too long, I would just as soon stay over till it is all done. Not that I wouldn't like to be home even if it were for a short time.

Since I last wrote to you, I have had quite an experience. I had a four day leave and went to Rome. They have an Army Rest Camp there that I went to. It didn't really pan out to be a rest at all as I was much more tired when I got back than when I went.

To begin with it was quite a thrill to have my first airplane ride. We went in a 1324 Bomber and it was OK. We passed over Anzio Beach Head on the way and saw in the fields below lots of bomb craters and wrecked tanks. We also went over Casino and saw what little was left of that. On the return we passed over Naples and Mt. Vesuvius. It was just my bad luck to be sitting in the flight deck on the wrong side of the plane and I didn't know we passed them. Would liked to have seen Vesuvius.

Boy that trip to Rome is sure worth a lot to a person. Wish everyone could get a chance to visit there. It has its good points and some not so good; as narrow streets and dirty stores. Of course you can't blame the people to be poorly dressed as they have had a strenuous time for some time. I actually saw dead sparrows in a shop; they were strung up in bunches of a hundred or so and were also laying on the counter. They must make pot pie out of them.

All the buildings, meaning the Historic ones are so large they are hard to describe. To begin with we saw the Pantheon which beside being built in 27 A.D. it is still in good shape. It has marble pillars in front weighing 80 tons each and these were brought from Egypt, 2000 miles over land and water, over 2000 years ago. We then went to Victor Emanuel 11 Memorial. It's a mammoth white structure covering 22,000 square yards. In the front, up I don't know how many stairs, is the Tomb Of The Unknown Soldier. Two guards stand there all the time, just as still as the statues. In front is also a statue of a large Bronze Horse with Victor Emanuel in the saddle. To look at it, it doesn't look so large, but upon completion the eighteen Architects held a banquet inside the Horse and still left room for the waiters to go between them.

From there we went to the Coliseum. It is a mammoth bowl, 158 feet high and at least a block in circumference. Part of the wall is still standing and we climbed up thru the corridors till we were near the top. We were told that when they had Gladiators fights in the olden days, they used to rig up an awning to complete the structure. To do this they had the sailors of the Imperial Navy pull a hemp loop up in the center and string riggings out from that. It was then covered by an embroidered cloth. Close by this was the ruins of the house of Nero. It was two miles long. Well I am going to take the time out and go to a USO Show that is about ready to start.

Well here it is in the evening of the next day. Went to the show last night and didn't feel at all like writing today. It was my day off and I am glad it was as I spent all the time in the sack. Sure wish I could get to feeling good. As I was saying, I went to the show last night and it was a good one. The name of the show was the LOST ANGEL. Also I have seen a couple of Dr. Gillespie shows that I thought were swell, HIS NEW ASSISTANT and THREE MEN IN WHITE. The USO show was put on by all GI Negroes and was well worth seeing.

Glad I didn't send yours or Nellie's letter last night as I received letters from each of you. Thanks a lot for the picture. It sure is swell of all of you. Shirley sure is growing up and looks very much like her Mother.

I don't believe I mentioned receiving your other letter, Helen? I received the one written on August 4th. It took a long time to get here, didn't it? Sure would have liked to help Harold on the Model A, or mostly on the quart of whiskey. They had a little impromptu celebration on the 4th in the line of shooting flares. They really lit up the sky. We have not heard of anything like a blackout since we left the states funny isn't it. We had our big fire works on the way over. Would like to tell about it but aren't allowed to. Never want to go through that again. Boy I am making a lot of mistakes. In your second letter you start out by saying, I've been doing swell in writing to you, but I don't think I've been doing so good. This letter should be a good time to give congratulations on your 16th anniversary, Helen and Harold. That's quite a few years and hope you have many more like them.

Well I had better get back to the tour of Rome. After seeing the Coliseum, we drove out to the Catacombs. Altogether in Rome there are 900 miles of burial tunnels and over 2,000,000 graves in them. Saw where St Peter is buried; also St. Paul. Next we saw the ruins of less interest such as The Circus Maximus, The Roman Forum Castle of San Angelo and Arch of Constantine. These were all built about the time of the birth of Christ; some before.

The highlight of the whole trip was the visit to the Vatican City and St. Peter's Cathedral. We went all through the grounds and then the Guide took us through the Cathedral. I never saw such a large place before. The Church is 696 feet long and 499 feet wide. From the floor to the dome is 195 feet. You just can't explain a place like that. You have to see it. It's too beautiful to try and tell about it. The top and sides are all covered with gold like designs and fancy work. The Mosaics on the wall are about 20 x 35 feet in size and look like oil paintings, but really are half inch pieces of different colored marble, a foot long set in on end. You can imagine how long it must have taken to make them.

There was a large Bronze Statue of St. Peter that when people passed, they would touch and kiss the toes. In the years that this has been done, one half of the toes are worn away. Another sight I didn't expect to see was the Pope. I bought a string of Rosary Beads for Bob Spain and as the Pope went by, I had them blessed. I think he will appreciate them.

Well I think I will cut this short now as I think that takes in most of the news. Will write a little sooner than I did last time but perhaps this one will make for it.

Beulah, Don it and Sue, I hope you will understand my reason for writing to you all at the same time. As is I am getting writer's cramps. (That is all there was end of story). FINITO

# UNIT HISTORY, 485TH BG (H) 1 JULY 1944 TO 31 JULY 1944

The 485th BG (H) flew 18 operational missions during the month, making a grand total of 51 missions as 31 July 1944. 1262 tons of bombs were dropped. Results were generally very good. Mission photos showed strikes on the target or in the target area on all but two missions

Eleven (11) aircraft and their crews were lost in combat operations.

The Group's score of enemy aircraft encounters was as follows:

20 Destroyed
5 Probables
3 Damaged

The Group's morale maintained its high level. An outstanding factor was the continuation of quotas of Officers and Enlisted Men for the various rest camps. A new one added this month, was welcomed by all - was a quota of Officers and Enlisted Men for weekly visits to Rome. The presentation of several USO shows was welcomed by all and enthusiastically received. Athletic activities organized by Special Services were always popular. Sports included softball, swimming and volleyball.

On 16 July 1944, formal dedication was made of the new Group Chapel, the only individual Group Chapel (solely for religious services) in the 15th Air Force.

Sixteen (16) replacement crews survived and were given the orientation and indoctrination program already in effect.

During the month, the DFC was awarded to the following personnel for meritorious achievement while participating in aerial flight:

Col. Walter E. Arnold, GP CO - 2 July
Major Edward H. Nett, 828th CO - 12 July
Major Roy L. Reeves, Asst. GP Off - 13 July
Lt. John N. Coffin, 831st Navigator - 13 July
Lt. Albert C. Hawklins, 828th Bombardier - 13 July
Lt. Lloyd F. Towers, 829th Navigator - 13 July
Lt. Frank T. Wodzinski, 829th Pilot - 13 July (later reported as MIA)

### Rest Camp

After the Aircrews had completed about one half of their missions they were sent to a Rest Camp for rest and relaxation (R&R). The Ground Crew also needed a break from the daily routine. Pop Arnold wanted to give each man an opportunity to visit Rome. This was quite a treat because, unlike the youth of today who can "backpack" all over the world, most of us never had the chance to travel because of the Depression. Some of us had not been more than 50 miles from home before entering the service. It also gave the ground crew an opportunity to ride in a B-24 which had become a big part of their daily life.

## MISSION NO. 34 - 2 JULY 1944

At 0658 hours, 37 B-24's took off to bomb the BUDAPEST RAKOS MARSHALLING YARD in Hungary. The 1st attack unit was led by Major Daniel L. Sjodin, 831st CO and the 2nd attack unit was led by 1st Lt. Francis P. Tunstall, 830th Flt Cmdr. One aircraft returned early. Fighter escort consisted of 25 to 30 P-51's and was effected at 0935 hours. The escort left the formation on return at 1130 hours. Aircraft 143, 493, 144 and 694 returned early.

Fifteen (15) enemy aircraft - Me 109's and FW 190's were seen between the IP and the target area but did not attack the formation. MAH flak was encountered over the target for approximately 5 minutes.

Thirty-two (32) aircraft were over the target at 1030 hours and dropped 79.5 tons of 500 lb. GP bombs at 22,400 feet. Bomb strikes were seen in the target area. Results were unobserved due to smoke which came up to 10,000 feet.

Ten (10) aircraft of our formation were holed by flak - damage slight. Two Me 109's were seen to go down in flames at 1030 hours in the target area after being attacked by fighter escort. There were no losses.

The weather was clear over the Adriatic except for surface haze layer. Large patches of cirrus stratus at 22,000 feet in the Budapest area and heavy, persistent condensation trails.

Return was made without incident, 32 aircraft landing at 1330 hours.

## MISSION NO. 35 - 3 JULY 1944

At 0815 hours, 32 B-24s took off to bomb the TIMISOARA MARSHALLING YARD at Timisoara, Rumania. Col. Walter E. Arnold, Jr., GP CO led the 1st attack unit and 1st Lt. John M. Jones, 831st Flt Cmdr led the 2nd attack unit. The Group rendezvous with the 460th BG. Rendezvous with the fighter escort, 20 to 30 P-38's, was effected at 1035 hours. The fighter escort left the formation on return at approximately 30 miles off the Yugoslavian Coast. Aircraft 729 returned early.

Two unidentified enemy aircraft were seen enroute. There was no encounter with enemy aircraft. Flak at the target was SIH.

Thirty-one (31) aircraft were over the target at 1121 hours and 30 aircraft dropped 74.5 tons of 500 lb. Cluster amiable Incendiary bombs from 21,000 feet. Bomb strikes were observed in the target area. A good pattern of bombs was made over the target. Heavy, black smoke rolled up to 6,000 feet over the target.
Weather: Broken clouds and good visibility.

Return was made without incident, 31 aircraft landing at 1,130 hours. There were no losses.

## MISSION NO. 36 - 5 JULY 1944

At 0755 hours, 33 B-24's, scheduled, took off to bomb the TOULON SUBMARINE DOCKS (PT) at Toulon, France. The 1st attack unit was led by Major Richard V. Griffin, 830th CO and the 2nd attack unit was led by Captain Thomas D. O'Brien, 829th CO. One aircraft, which dropped its bomb load accidentally, returned early - before assembly. The Group rendezvous with the 460th BG, 464th BG and 465th BG. Rendezvous with the fighter escort, consisting of 1 Group of P-38's, was accomplished at 1230 hours. The fighter escort remained with the formation over the target and until the formation was well past the coast of France on the return route. One aircraft returned early, prior to bombing.

Twenty-nine (29) aircraft were over the target at 1239 hours and dropped 72.5 tons of 500 lb. GP bombs from 22,800 feet. Two aircraft are missing. Result's of the bombing was generally fair. The enemy attempted to use a smoke screen to obscure the target which was only partially successful due to an adverse wind. No hits were observed on the submarine pen, however, some bombs were seen to hit the adjacent dry docks.

No enemy aircraft were encountered. Two enemy aircraft were seen in the target area and were engaged by the P-38 fighter escort. Moderate to intense - accurate - heavy flak was encountered over the target. Several aircraft received minor damage from the flak. Some red flak bursts were also observed at the target area. The Nose Turret Gunner of aircraft 727 had a very exciting experience just before the target was reached, his electrical flying suit caught fire. It was extinguished without much difficulty, the gunner receiving only, slight burns.
Two Aircraft are missing:

Aircraft 812 was observed to have turned back at 1200 hours with no. 3 engine feathered. When last seen, the aircraft was under control and heading towards Corsica.

Aircraft 127 was observed to lose altitude at the target and to lag behind the formation. Several P-38's were seen to drop down to provide cover.
Aircraft was last seen just past the target at approximately 1241 hours, at which time it was under control. The weather was clear.
Return was made without incident, 29 aircraft landing at 1515 hours.

## MISSION NO. 37 - 6 JULY 1944

At 0712 hours, 32 B-24's took off to bomb the PORTO MARGHERA OIL STORAGE (PT) near Mestre, Italy. The 1st attack unit was led by Major Robert E. Smith, GP Ops Officer and the 2nd attack unit was led by Captain Ralph E. Monroe, 828th Ops Officer. The Group rendezvous with the 460th Group. Rendezvous with the fighter escort - consisting of 19 P-38's was effected at 0929 hours. The escort left the formation at the target. There were no early returns.

Thirty-two (32) aircraft were over the target at 1037 hours and dropped 80 tons of 500 lb. GP bombs from 20,600 feet. Bomb strikes were seen in the target area and some bombs were seen to go over. Smoke and fires were observed over the target.

Three Me 109's were seen in the target area, however there were no encounters. Twenty (20) aircraft, of the formation were holed by flak. Damage was slight. There were no losses. The weather was clear over the Adriatic.

Thirty-two (32) aircraft landed at base at 1304 hours without incident.

## MISSION NO. 38 - 7 JULY 1944

At 0634 hours, 32 B-24's (36 scheduled) took off to bomb the BLECHHAMMER NORTH SYNTHETIC OIL PLANT (PT) in Germany. The 1st attack unit was led by Major E. H. Nett, 828th CO and. the 2nd attack unit was led by Captain James F. Hogan, 831st Flt Cmdr. Aircraft 096, 536, 728 and 494 failed to take off. Aircraft 498 returned before rendezvous. The Group rendezvoued with the 460th BG, 464th BG, and the 465th BG. Rendezvous with the fighter escort - 20 P-38's - was effected at 0905 hours. Escort departed the formation on return at 1320 hours. Aircraft 132, 422 and 699 returned prior to bombing.

Twenty-eight (28) aircraft were over the target at 1119 hours and dropped 67.5 tons of 500 lb. GP bombs. Bomb strikes were seen in the target area. Smoke screen and heavy smoke from the target obscured definite results. Black smoke over the target was seen up to 16,000 feet. Flak at the target was Ml to AH.

Twelve (12) enemy aircraft were seen - 3 Me 109's in the target area, 3 FW 190's and 6 Me 210's in the Lake Balaton area. None of the enemy aircraft attacked the formation. Five (5) aircraft of the formation were holed by flak. Damage was slight. Flak at the target appeared to be rocket flak as there were large red balls of fire coming up before exploding. There were no losses.
The weather was good.
Twenty-eight (28) aircraft landed at 1451 hours.

## MISSION NO. 39 - 8 JULY 1944

At 0633 hours, 29 B-24's (31 scheduled) took off to bomb the FLORISDORF OIL REFINERY (PT) at Vienna, Austria. The 1st attack unit was led by Lt. Col. William L. Herblin, Deputy GP CO, and the 2nd attack unit was led by Captain Roy L. Reeve, Asst. GP Ops Officer. Aircraft 414 and 536 failed to take off. The Group rendezvoued with 460th BG, 464th BG and the 465 BG. Rendezvous with the fighter escort 25 P-51's - was effected at 1000 hours. Escort left the formation at 1232 hours. One aircraft returned early.

Twenty-eight (28) aircraft were over the target at 1040 hours and dropped 68.75 tons of 500 lb. GP Bombs. Bombs were seen to strike short of the target and in the target, flames were seen to leap up to 5,000 feet with black smoke up to 10,000 feet.

About 30 enemy aircraft were seen between the target and Papa, of which 20 were Me 109's and 10 FW 190's. One FW 190 attacked the aircraft in no. 2 position in the low box of the 2nd unit. Attack was made from 12 o'clock low and was fired upon by the ball turret gunner at 1042 hours in the target area. This enemy craft is claimed as a possible. The enemy aircraft broke off at 300 yards out of control. Color of the enemy aircraft was brown with green camouflage.

Eighteen (18) aircraft of our formation were holed by flak. Damage was slight. Co-Pilot of no. 1 aircraft, low box of the 1st unit was slightly injured by flak. Three of our aircraft were lost, apparently to flak in the target area. Aircraft 674 was hit in the wing and seen to go down at 1040 hours over the target. Seven chutes were seen from this aircraft. Aircraft 769 was seen to go down at 1105 hours and 10 chutes were seen to open. Aircraft 1164 was seen to go down at 1101 hours with two engines out. Ten chutes were seen to come out of this aircraft. Smoke screen was observed 2 miles NW of the Primary target.

At 1040 hours, aircraft 674 received a direct hit by flak over the target.

No 4 engine was knocked off. The aircraft then went into a spin with the right wing burning. Seven chutes were seen to open close to the ground.

At 1105 hours, aircraft 769 was seen to go down with two engines smoking. Ten chutes were seen to open.

At 1101 hours - aircraft 1164 went down. Ten chutes were seen to open. Lt. Edwin H. Sibila, in aircraft 418, followed aircraft 1164 as it dropped out of formation. Two engines were cut and the crew was seen to be changing clothes. Then they bailed out. The aircraft appeared to be on A-5, as it flew a steady course for a few minutes before it crashed.

The weather was clear over the entire route.

Twenty-five (25) landed at base at 1329 hours.

## MISSION NO. 40 - 12 JULY 1944

At 0606 hours, 39 B-24's, scheduled, and took off to bomb the NIMES MARSHALLING YARD (PT) at Nimes, France. The 1st attack unit was led by Col. Walter E. Arnold, GP CO and the 2nd attack unit was led by Captain Ralph E. Monroe, 828th Ops Officer. Wing rendezvous over Spinazzola was not satisfactory. The order of flight as briefed was 465th BG, 464th BG, 460th BG and 485th BG. The formation arrived over Spinazzola at 0731 hours. However, no other Groups were seen. The 485th continued on course and it was not until the formation was well over the water that the other Groups joined. Due to the mix-up at the rendezvous, the order of flight was varied, inasmuch as the 485th went over the target in the 3rd position. There was no fighter escort and no early returns.

Thirty-nine (39) aircraft were over the target at 1133 hours and dropped 97.5 tons of 500 lb. GP bombs. The target was well covered by a concentration of bombs. The locomotive works, the car repair shops, a warehouse area, a choke point and railway sidings, with many goods wagons, received direct hits. In addition, several strikes were observed in a barracks's area north of the car repair shops.

MAH flak was encountered near Beaucaire. Eight (8) aircraft of the 2nd unit received minor damage. Formation was exposed intermittently for about 5 - 10 minutes of flak. Over the target, the formation encountered scant flak.

Fifteen (15) enemy aircraft were seen in the vicinity of the target. Two of these, ME 109's, flew at the same altitude as the formation, just out of range, for 8 to 10 minutes, joining before the IP was reached and continuing beyond. It is believed that they did this in order to radio the altitude of the formation to the anti-aircraft defenses. The enemy aircraft appeared to concentrate their attacks on the Group ahead. Some rocket fire from enemy aircraft was observed.

Enroute at 0915 hours, aircraft 750 was fired on by an unidentified friendly bomber, presumably while test firing its guns. Aircraft 750 received several hits; the glass in the nose turret being demolished; two holes in the pilot's windshield; one hole in the right wing and one hole in the tail. This caused excitement among the crew members, but the pilot continued with the formation to the target. It is believed that bullets came from another Group, since this Group did not check fire its guns due to the proximity of other Groups.

Weather: There was an overcast at the base at take off time. Over the target, clear weather existed. Due to the weather conditions, aircraft returned to the base individually or by boxes. Thirty-one (31) aircraft landed without incident at 1507 hours.

Several aircraft landed at friendly fields.
Aircraft 694 landed at Corsica.
Aircraft 121, 730, 440, and 110 landed at Naples.
Aircraft 729 landed at Naples
Aircraft 495 and 143 landed at the 460th BG air field.

## MISSION NO. 41 - 13 JULY 1944

At 0717 hours, 32 B-24's (33 scheduled) took off to bomb PORTO MARGHERA (PT) in northern Italy. The 1st attack unit was led by Major Daniel L. Sjodin, 831st CO and the 2nd attack unit was led by Major John B. Stoddart, 830th Ops Officer. Col. George G. Acheson, 55th Bomb Wing CO rode with Major Sjodin. The Group rendezvous into wing formation, and with fighter escort - 12 P-38's - at 1015 hours. The escort left the formation at 1057 hours.

Thirty-one (31) aircraft were over the target at 1032 hours and dropped 76.25 tons of 500 lb. GP bombs. Preliminary assessment on photos show approximately 15 bomb strikes in the north half of the target area; some strikes are shown to have damaged the factory area located southwest of the primary target. Several fires were observed in the target area.

Three Me 109's were seen in the target area, however, none attacked the formation
Flak was encountered over the target area for approximately 4 minutes.
Eleven (11) aircraft of our formation were holed by flak. Damage was slight.
The weather was partly cloudy.
Thirty-one (31) aircraft returned safely, landing at 1243 hours. There were no losses.

## MISSION No. 42 - 14 JULY 1944

At 0617 hours, 36 B-24's (38 scheduled) took off to bomb the OIL STORAGE at PORTO MARGHERA (PT) in northern Italy. The 1st attack unit was led by Major R. V. Griffin, 830th CO and the 2nd attack unit was led by Captain Thomas D. O'Brien, 829th CO. Last resort target - MARSHALLING YARD at MANTUA, Italy - was bombed because primary and alternate targets were closed by 10/10 undercast. The Group assembled into Wing formation. Rendezvous with the fighter escort - 20 to 25 P-51's - was effected at 0925 hours. The escort left the formation at 1012 hours. Aircraft 164 and 089 returned early. No enemy aircraft or flak encountered. Thirty-four (34) aircraft were over the target at 0945 hours and dropped 84 tons of 500 lb. Cluster Incendiary bombs from 21,000 feet. Bombs were seen to hit short of the target and some hits were observed in the target area in the Marshalling yard. The weather was clear over the base and cloudy in the target area.

Thirty-four (34) aircraft landed at 1259 hours.

## MISSION NO. 43 - 15 JULY 1944

At 0647 hours, 37 B-24's (38 scheduled) took off to bomb the UNIREA SPERANTZA OIL REFINERY in Ploesti, Rumania. The 1st attack unit was led by Lt. Col. William L. Herblin, Deputy GP CO and the 2nd attack unit was led by Major John E. Atkinson, 831st Ops Officer.

The Group assembled into Wing formation. Rendezvous with the fighter escort - 40 P-51's - was effected at 1020 hours. Twelve (12) P-38's were observed over the target area. Fighter escort departed from the formation at 1128 hours.

Aircraft 724, 416, 857 and 601 returned early. Lt. Col. Herblin in aircraft 857 left the formation due to engine malfunction. Deputy leader 1st Lt. Ray M. Smith, Jr., of the 828th Sqdn, took over the lead and proceeded to the Primary Target.

Thirty-two (32) aircraft were over the target at 1058 hours and 29 aircraft dropped 71.5 tons of 1000 lb. GP bombs. The bombs were dropped in a smoke covered area and through and overcast. No results were observed. Smoke could be seen on return for a distance of 140 miles. Photos revealed the target to have been completely smoke covered.

No enemy aircraft were encountered. Flak was encountered at the target for 5 minutes and was IAH. Eleven (11) aircraft received minor damage from flak.

Weather: There were a few altocumulus clouds over the base. There was 8/1- cloud coverage over the target.

Thirty-two (32) aircraft landed at 1434 hours. There were no losses.

## MISSION NO. 44 - 16 JULY 1944

Aircraft plant - Wiener Neustadt, Austria — Mainly for Manufacturing engines for ME-109's.

NO RECORD RE ABOVE MISSION IN HISTORY

## MISSION NO. 45 - 19 JULY 1944

At 0645 hours, 36 B-24's took off to bomb the ALLACH BMW MOTOR WORKS (PT) in Munich, Germany. The 1st attack unit was led by Major Nett, 828th CO and the 2nd attack unit was led by Major John E. Atkinson, 831st Ops Officer. The Group assembled into Wing formation. Rendezvous with P-38's and P-51's of the 306th FW was effected at 1015 hours. The escort left the formation at 1235 hours.

Twenty-nine (29) aircraft were over the target at 1144 hours and 28 aircraft dropped 70 tons of 500 lb. GP bombs. The primary target was covered with a very effective smoke screen, in addition to 2 to 3/10 cloud cover. It being necessary to use navigational aids due to obscured target. The lead ship was handicapped in determining the amount of drift approaching the target from the IP. The Operator realized too late that he had to make a 10-degree drift correction due to wind changes which materially effected the pre-set of 3.5 left drift, when only 4.5 miles from the primary target, and traveling at a ground speed of 250 mph, which also had not been predicted. As a result, the bombs over shot the target. Hits were seen, however, on a small marshalling yard and probable military stores.

Flak at the target was IAH. The formation was subjected to flak for 6 to 7 minutes, in spite of the fact that ground speed was over 250 mph. Three chutes were observed to open. Part of the aircraft's wing flew off, damaging the top turret gunner. Several aircraft sustained minor to moderate flak damage and one Navigator was slightly injured by flak. About 10 - 15 enemy fighters, probably Me 109's were seen in the target area and did not attack the formation.

One aircraft of the formation, 444, YG was lost to flak at 1145 hours. Flak hit the no 2 engine, blowing the left wing off the aircraft. Three chutes were seen to open. Two minor casualties were sustained: one by the top turret gunner and one by a Navigator.

Weather: There were 4/10 cumulus over the base and 2 to 3/10 altocumulus over the target. Twenty-eight (28) returned to base at 1455 hours.

## MISSION No. 46 - 20 JULY 1944

At 0650 hours, 37 B-24's, scheduled, took off to bomb the LUFSCHIFFBAU ZEPPLIN WORKS at Friedrichschafen, Germany (PT). The 1st attack unit was led by Major Sjodin, 831st CO and the 2nd attack unit was led by Captain William D. Jernigan, Jr., 830th "F" Flt Cmdr. The Group rendezvous with the 464th and the 465th Bomb Groups. Rendezvous with P-51's, which provided penetration cover to the target, was effected at 1025 hours. P-38's were picked up over the target and provided target and withdrawal cover as far back on the route as the Adriatic Coast.

Three aircraft returned prior to bombing. One aircraft returned shortly after take off. Aircraft 495 lost an aileron after another B-24, in the formation, (361), having been hit by fighters and partially out of control, crashed into the aircraft, the wing of 361 knocking the aileron off aircraft 495. The pilot and co-pilot of aircraft 361 - which spun and crashed - deserve commendation for averting a more serious crash between the two aircraft. Before returning to base, aircraft 495 dropped 5 bombs on a bridge - target of opportunity - at 1001 hours from 17,000 feet with unobserved results. The 3rd aircraft, 299, having had one engine shot out by fighters, nevertheless, continued on with the formation almost as far as the IP when another engine developed trouble, compelling the aircraft to turn back. This aircraft jettison 5 bombs to lighten the load and returned to base. About 15 enemy aircraft appeared at 0956 hours in the Udine (Italy) area, probably all Me 109's and FW 190's. Aggressive attacks were made at the high box of the 1st unit and the high box of the 2nd unit by at least 6 Me 109's and FW 190's. The Me l09's attacked mainly from 10 to 1 o'clock high singly and the FW 190's attacked in threes and fours, high from the rear and the front. It is believed that rocket shells were used by some of the enemy fighters, as large red bursts were seen as the enemy aircraft attacked. Two B-24's of the formation were shot down by enemy aircraft. Our gunners claimed 2 FW 190's destroyed and 2 Me 109Ps destroyed and 2 Me 109's damaged. The enemy aircraft, were for the most part, black with silver markings and one crew reported two Me 109's with red circles painted on the bottom of the wings, which were dark camouflage color.

Thirty-two (32) aircraft were over the target at 1055 hours and dropped 79.5 tons of 1,000 lb. GP bombs from 23,800 feet. Bombs were seen to fall in the target area and a marshalling yard near the target received several direct hits in the choke point. The whole target area was observed to be covered with smoke as the formation left the target.

Flak over the target was IAH and the formation was in the flak for 5 minutes. Aircraft 722 was lost to flak over the target. Several injuries, due to flak, were incurred and one Navigator was killed by flak. No enemy aircraft were identified in the target area and there were no encounters.

Three (3) aircraft were lost - two to enemy aircraft and one to flak. The two aircraft, which were shot down by enemy aircraft, 361 and 88, a total of 17 chutes were seen from them. The third aircraft 722 was shot down by flak over the target and at least 5, chutes were seen to open. Aircraft 713 incurred severe battle damage and was left at Panatella Field and the crew returned to base at 1624 hours.

There were 5 casualties. In one aircraft, which was badly holed by flak, the Navigator was killed and the engineer and the ball turret gunner were wounded.
Another Navigator sustained a broken arm from, flak and was landed at Foggia, (Italy) to be treated. The Bombardier of the lead aircraft received a hand injury from flak on the bomb run.

Weather: It was clear over the target with a slight haze. Enroute there was less than 1/10 scattered altocumulus clouds.

Twenty-nine (29) aircraft returned to base at 1425 hours. One aircraft landed with a flat right tire and the last aircraft landed with practically no hydraulic fluid as the hydraulic system had been shot out by flak. Two aircraft landed at Panatella Field. Aircraft 713 had injured personnel aboard and was so badly damaged that it was left at the airfield. The other aircraft 818, because of fuel shortage, returned to base at 1624 hours.

## MISSION NO. 47 - 22 JULY 1944

At 0806 hours, 37 B-24's took off to bomb the ROMANO AMERICAN OIL REFINERY (PT) at Ploesti, Rumania. The 1st attack unit was led by Col. Arnold, GP CO and the 2nd attack unit was led by Lt. Ray E. Smith, 828th "A" Flt Cmdr. The Group assembled into Wing formation. No rendezvous was made with the fighter escort enroute to the target. Some P-51's and P-38's were seen over Yugoslavia on the return - about 25 in all. Fourteen (14) aircraft returned early of which 11 of these aircraft landed at Pantanella Field because they could not land with the strong cross wind, which prevailed at the home field.

Twenty-three (23) aircraft were over the target at 1140 hours and 21 aircraft dropped 51.5 tons of 1,000 lb. GP bombs from 24,000 feet. Bomb run was made on C-1. Disposition of bombs of two other aircraft is not known. One of these aircraft landed at a friendly field and the other has been reported at the island of Vis. Results of bombing could not be definitely determined because of and undercast and a combination smoke screen and black oil smoke. Flak at the target was for the most part, intense-accurate-heavy. The formation was exposed to flak for about 4 minutes. Thirteen aircraft were holed by flak, none seriously.

There was one minor casualty due to flak. The enemy opposition consisted of one Me 109 seen over the target which did not attack and one FW 190 which came in to within 1,500 yards of the formation shortly past the target and then peeled off and left. Bombing was done strictly by PFF.

Weather: The target was mostly obscured when the formation passed over it. Twenty-one (21) aircraft returned to home field in spite of a strong cross wind, which prevented some of the early returns from landing at this field. One aircraft, which had incurred some flak damage and did not want to risk the crosswind, landed at Pantanella Field. The last aircraft was reported at 2230 hours to have landed at Vis. Aircraft landed at home base at 1546 hours.

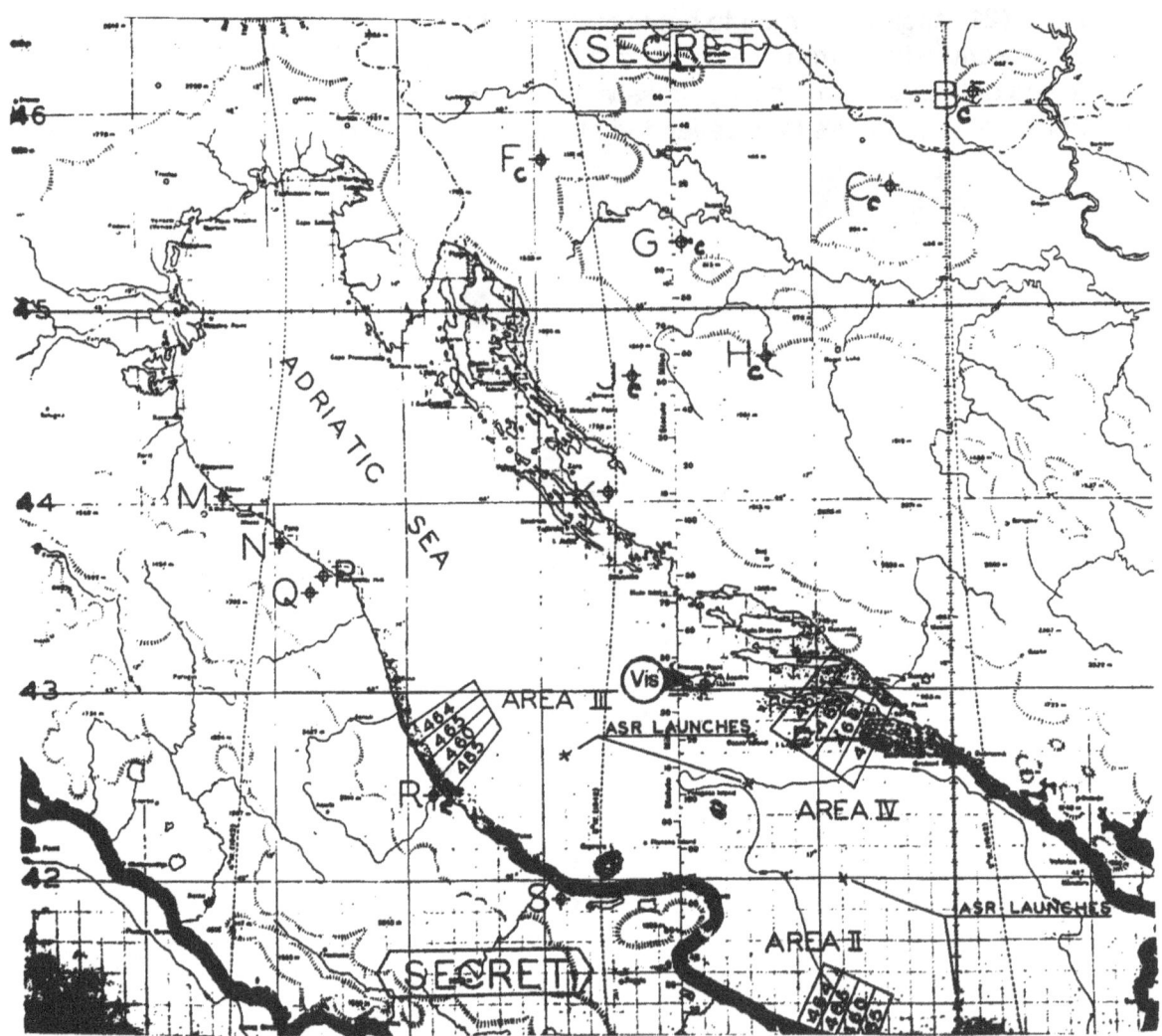

**EMERGENCY LANDING STRIPS**

The letters "A to S" on the above map shows the location of secret air strips in Yugoslavia, Hungry and Northern Italy which could be used in an emergency. (A list of coordinates was also included). Most of the air strips were short and camouflaged which made them hard to find. I think most crews would have elected to bail out rather than risk a crash landing. We were told that we would get help from the local people who were sympathetic to our cause.

Yugoslavia was in the middle of a civil war between the Chetnicks and the Partisans; however, both sides were friendly to us and helped many crews return to Italy. If we had to land there we were told not to discuss politics or be intimate with the women. One crewman who returned described the women. "They were rough, strong and husky. They carried guns and fought along side of the men. One woman surprised me with a big hug and picked me up off the ground". The Partisans led by Tito eventually won the civil war and executed the Chetnik leader despite pleas of leniency from crews who had received help from the Chetniks.

Also shown on the map are locations of Air-sea Rescue Launches on the Adriatic Sea that were available to pick up crews that were forced to ditch.

## MISSION NO. 48 - 24 JULY 1944

At 0705 hours, 35 B-24's took off to bomb the BALENCE/LA TIESURERIE A/D at Valence, France. The 1st attack unit was led by Col. Herblin, Deputy GP CO and the 2nd attack unit was led by Major John E. Atkinson, 831st Ops Officer. The Group assembled into Wing formation. Rendezvous with the fighter escort - P-51's of the 306th FW - was effected at 1100 hours, which provided penetration, target and withdrawal cover back to the French coast.

Thirty-four aircraft were over the target at 1151 hours and dropped 60.5 tons of 100 lb. fragmentation bombs. Two definite patterns of bombs were observed, one covering the hangar facilities and the other covering part of the landing ground and at least two aircraft revetments. No estimate of damage to enemy aircraft could be made because of partial cloud coverage.

Four Me 109's were identified in the distance between the target and the French coast on return at 1230 hours. None of these enemy aircraft made any attacks. Scant - moderate, accurate - heavy flak was encountered. One aircraft was moderately damaged due to flak and one engine had to be feathered. Several other aircraft received minor flak damage. There were no casualties. The tail of one aircraft was slightly damaged due to the explosion of a fragment cluster about 100 feet below the aircraft over the target area.

Weather: Almost clear over the Med. Sea. There were altostratus clouds at 12,000 feet in the vicinity of the target.

Thirty-four (34) aircraft landed at base at 1525 hours.

## MISSION NO. 49 - 25 JULY 1944

At 01704 hours, 35 B-24's (36 scheduled) took off to bomb the LINZ/HERMAN GORING TANK WORKS (PT) at Linz, Austria. The 1st attack unit was led by Lt. Col. R. E. Smith, GP Ops officer and the 2nd attack unit was led by Captain George W. Tompkins, 830th Flt Cmdr. The Group assembled into wing formation. Rendezvous with several P-38's was effected at 0930 hours, which covered the formation on penetration. At 1015 hours, 30 P-51's joined and furnished target and withdrawal cover and left the formation at the coast of Yugoslavia. Aircraft 422, and 834 returned early.

Approximately 11 enemy aircraft were seen over the target. Seven of these were Me 109's and four were FW 190's. They did not approach the formation. Several vapor trails were seen high over the target and at some distance. These believed to have been the results of encounters between enemy aircraft and fighter cover. Enroute, SIH flak was encountered near Zagreb.

Thirty-two (32) aircraft were over the target at 1129 hours and dropped 80 tons of 1,000 lb. GP bombs from 24,000 feet. Being unable to locate the IP, the bomb run was be run on ETA. The target was well covered by smoke on the approach.
This was caused by bombing of other Groups and from smoke pots surrounding the target. It appeared that the smoke pots were started only a few minutes prior to the arrival of the formation, indicating that the enemy was surprised. The wind tended to blow the smoke away from the target. The lead bombardier, unable to see his aiming point, bombed by offset, locating a bridge to the north and a stream intersection to the southwest. His bombs were

observed to have fallen into the smoke covered are believed to be the target. The smoke somewhat dissipated when the 2nd unit came over the target, leaving a portion of it in the clear. One string of bombs was observed to have hit the ore thawing plant, northeast corner of the target area. Smoke was observed coming from the machine and assembly shop in the east part of the target area. It is presumed that the 1st unit's bombs fell in that general area.

No enemy aircraft were encountered. Flak at the target was IAH. To the left of the target area, a barrage was thrown up. Burst at the formation indicated that the enemy was tracking. In addition to the above, several large white bursts were observed about 1,000 feet over the formation. Fifteen (15) of our aircraft received light flak damage. There were no casualties.

Weather: It was clear over the base and 5/10 altocumulus in the target area but clear over the immediate vicinity of the target.

Thirty-one (31) aircraft landed at base at 1414 hours. Aircraft 344 landed at Spinazzola, Italy.

## MISSION NO. 50 - 26 JULY 1944

At 0727 hours, 34 B-24's took off to bomb the ZWOLFAXING AIRDROME INSTALLATIONS and dispersed aircraft near Vienna, Austria (PT) and the SZOMBATHELY AIRDROME INSTALLATIONS at Szombathely, Hungary (1st alternate target. The 1st attack unit was led by Major Richard V. Griffin, 830th CO and the 2nd attack unit was led by Major John E. Atkinson, 831st Ops Officer. Aircraft 164 and 694 returned early. The Group rendezvous with the 460th Bomb Group. Rendezvous with fighter escort - 35 P-51's - was effected at 1034 hours and the escort left the formation at 1230 hours. Aircraft 494 returned prior to bombing.

Thirty-one (31) aircraft were over the target at 1129 hours and 7 aircraft of the low box, lead wave dropped 17.5 tons of 500 lb. GP bombs on the primary target from 20,500 feet. Results were poor. The formation proceeded to the alternate target and 20 aircraft dropped 50 tons of 500 lb. GP bombs at 1147 hours.
Results were poor. About 35 to 40 Me 109's and FW 190's were seen between Feldbach and Muryusclag and 10 Me 109's were seen in the Zagreb, Yugoslavia area on return. Approximately 30 of these enemy aircraft made passes at the formation. In the first attack, there were about 15 Me 109's flying high and to the right, parallel with the formation. They then peeled off and came in on the 2nd unit between 12 and 2 o'clock high, 2 or 3 abreast and 3 or 4 staggered slightly. They broke away low and attacked the formation from 5 to 6 o'clock, level and low. Our bombers claimed 5 FW 190's destroyed, 7 Me 109's destroyed, 2 FW 190's probably destroyed, 2 Me 109's probably destroyed, 4 FW 190's and 3 Me 109's damaged for a total of 23 claims. Markings on enemy aircraft: FW 190's - light grayish blue color, black crosses outlined in white with silver undersides. Flak at the alternate target was SIH. One aircraft of the formation was holed by flak, damage slight. Three aircraft were badly damaged by cannon fire and machine gun fire. The navigator on aircraft 438 was killed by a 20mm shell.

Weather was generally clear over the base and the Adriatic Sea. The target was cloud covered - 8 to 9/10 covered.

Twenty-nine aircraft landed at base at 1512 hours. Aircraft 438 returned late and aircraft 727 landed at a friendly airfield.

## MISSION NO. 51 - 26 JULY 1944

At 0636 hours, 35 B-24's (36 scheduled) took off to bomb the PLOESTI STANDARD OIL REFINERY (PT) at Ploesti, Rumania. The 1st attack unit was led by Major Ralph E. Monroe, 828th Ops Officer and the 2nd attack unit was led by Captain Francis X. Dalton, 829th "F" Flt Cmdr. Aircraft 725 returned before assembly. The Group rendezvous with the 464th BG and the 465th BG. Rendezvous with fighter escort - 30 P-51's - was effected at 0915 hours. The escort left the formation at 1237 hours. Seven (7) aircraft returned prior to bombing: 089, 394, 721, 699, 718 and 144. Twenty-seven (27) aircraft were over the target at 1020 hours and 26 aircraft dropped 65 tons of 500 lb. GP bombs. Bombing was done by PFF. Results were obscured by smoke screen over the target. Black smoke was rising up to 15,000 feet. Photos revealed strikes on course but over the target.

Six enemy aircraft, Me 109's, were seen over the primary target. None attacked the formation. Flak at the PT was IAH. It appeared to be extremely accurate. Thirteen (13) aircraft were holed by flak - damage was slight. The tail gunner of 699 was slightly wounded by flak. One of the crews observed a B-24 in the high box of the 464th BG drop its bombs on a B-24 in the low box of that Group during the bomb run, causing the aircraft in the low box to explode and go down.

Aircraft 330 ran out of fuel, losing all four engines. It was forced to crash land 3.5 miles southeast of the field. Pilot, Captain Marsden G. Kelly, landed the aircraft in a wheat field, landing gear down, causing minor damage to the aircraft. No crewmembers were injured.

Weather: Few scattered patches of altocumulus over the base. The target area was clear.

Twenty-five (25) aircraft landed at home field at 1145 hours. One aircraft landed at a friendly field.

## MISSION NO. 52 - 30 JULY 1944

Thirty-three (33) B-24's (34 scheduled) took off to bomb the BUDAPEST/DUNA AIRDROME INISTALLATIONS (PT) at Budapest, Hungary. The 1st attack unit was led by Lt. Col. Herblin, Deputy GP CO and the 2nd attack unit was led by Captain George W. Tompkins, 830th Flt Cmdr. The Group assembled into Wing formation. Rendezvous with fighter escort - approximately 75 to 90 P-51's and P-38's - was effected at 1032 hours. Six aircraft returned prior to bombing: RP, YC, YE, YW, BC, AND BA.

Twenty-six (26) aircraft were over the target at 1114 hours and 25 aircraft dropped 65 tons of 500 lb. RDX bombs from 24,000 feet. Bombing was accomplished under very difficult conditions. The initial point and the last two turning points were obscured by clouds. The formation turned short of the IP, on ETA, flying over 10/10 clouds. The target was picked up through a break in the clouds. During the bomb run the formation was flying directly into the sun, the glare on the haze making it very difficult for the bombardier to locate the target. The briefed aiming point was intermittently covered by scattered clouds and smoke on the bomb run. However, bombing was accomplished visually. Crew members reported seeing bomb strikes in the general target area, but the consensus of opinion seems that most bombs fell to the left and short of the target. Preliminary damage assessment shows that bombs fell short and to the left. No military damage was noted.

Approximately 9 Me 109's were in the target area. None of these attacked the formation. Fighter escort engaged the enemy aircraft after they had attacked another Group. Flak over the IP was SIH, lasting for 2 minutes. Over the target IAH flak was encountered.

    Two aircraft were lost to flak. On the bomb run, aircraft YU received a direct hit in the bomb bay. The aircraft turned over and it's bombs scattered. The aircraft dropped several thousand feet and broke in half, at which time three chutes were seen to emerge from the tail section. Aircraft WO released its bombs over the target and was hit at 1145 hours. One engine appeared to have been knocked completely off. The aircraft went down slowly, apparently under control. Six chutes were definitely seen to open with a good possibility that others got out before the aircraft crashed.

    Crewmembers reported that flak was very accurate and that it was of the directed - tracking type. The formation was in this flak for 5 minutes. On the return, the formation deviated slightly to the left of the briefed course to dodge a heavy cloudbank. Flak was encountered 6 miles north of Mostar and it appeared to be IAM. Since the formation was at 12,000 to 14,000 feet at this time, this flak was probably from automatic weapons. The formation was in this flak for approximately 2 minutes. Crews thought the flak came from along the highway.

    Weather: There was 1/1 - cumulus clouds over the base, and 5110 clouds over the target.

    Twenty-five (25) aircraft landed at base at 1355 hours.

# UNIT HISTORY, 485TH BG (H) 1 AUGUST 1944 TO 31 AUGUST 1944

The 485th BG (H) flew 20 operational missions during the month; making a grand total of 72 missions as of 31 August 1944. 1178.32 tons of bombs were dropped on numerous targets. Combat photos showed that the results were very good.

Five (5) aircraft and their crews were lost in combat operations. Col. Walter E. Arnold, Jr., GP CO, Lt. Col. Robert A. Smith, GP Ops Officer, Captain Kenneth. N. Gillespie, Group Bombardier, and Captain Jay (NMI) Jaynes, Group Navigator, were reported missing in action on 27 August 1944. The following men bailed out over Yugoslavia and were evacuated:

2nd Lt. Elmer B. Hall
2nd Lt. Clifford J. Manderscheid
2nd Lt. William H. Donovan
SSgt Kenneth S. Robinson
TSgt Edgar J. Helms, Jr.
Sgt. Richard M. Salazar
SSgt Forrest L. Yeager
Sgt. Thomas West
Sgt. Billy R. Culver

The Group's score of enemy aircraft encountered for the month was:
9 Destroyed
3 Probables
3 Damaged

The morale of the Group continued to maintain its high standard. Service Clubs have been opened in all of the Squadrons for Officers and Enlisted Men. Many of the men have built stone or wooden houses and showers have been constructed. Many improvements were made on the theatre.

During the month of August, a large number of our original crews completed their 50 missions and were scheduled to return to the States as soon as arrangements were made.

Nineteen (19) crews were assigned to this Group and were given a six-day program of indoctrination and orientation.

The following personnel were awarded the Silver Star:
2nd Lt. Hudson J. Owen - 19 August 1944
SSgt Paul M. Combs - 19 August 1944 (Reported missing in action 9 June 1944)
The following personnel were awarded the DFC:
Lt. Col. John P. Tomhave, GP CO - 28 August 1944
Lt. Col. William L. Herblin, Deputy GP CO - 8 August 1944
Captain Howard N. Cherry - 8 August 1944
Captain John D. Hanson - 8 August 1944
Captain William D. Jerningan - 8, August 1944 (Reported MIA, 20 July 1944)
Captain Roger J. Jones - 8 August 1944
2nd Lt. Clifford A. Martin - 8, August 1944
1st Lt. Robert B. Skelton - 8 August 1944
1st Lt. William M. Roberts - 11 August 1944
Major Ralph E. Monroe - 15 August 1944
1st Lt. Ray M. Smith, Jr. 15 August 1944

1st Lt. Joseph B. McNamara - 15 August 1944
1st Lt. Clement J. Hess - 15 August 1944
1st Lt. Jesse I. Ledbetter - 18 August 1944
Captain Jacob S. Disston - 20 August 1944
1st Lt. Albert E. Liddicoat - 20 August 1944
2nd Lt. Thomas E. More - 20 August 1944 (Reported MIA, 20 July 1944)
1st, Lt. William R. Boling - 24 August 1044

## MISSION NO. 53 - 2 AUGUST 1944

At 0915 hours, 15 B-24's, scheduled, took off to bomb the GENOA HARBOR INSTALLATIONS (PT) in northern Italy. The 1st attack unit was led by Major Roy L. Reeve, Asst. GP Ops Officer and the 2nd attack unit was led by Major Ralph E. - Monroe, 828th Ops Officer. The Group assembled into Wing formation. Rendezvous with fighter escort - several P-38's were effected at 1238 hours. The escort left the formation on return at 1345 hours. Aircraft 274 and 089 returned early.

Thirty-one (31) aircraft were over the target at 1301 hours and dropped 75.75 tons of 500 lb. GP bombs from 21,900 feet. Results: Black smoke was observed to 5,000 feet. Preliminary photo interpretation report shows four strikes on Harbor wall immediately south of the target area. It appears to be badly damaged in two places. Eleven strikes in the marshalling yard in the target area, causing damage to 6 to 12 wagons. Three strikes in the storage area and loading docks to the south side of target area. Two fires appear to have been started. Additional strikes believed to be in the marshalling yard and storage area but cannot be seen because of smoke cover from bombing of another Group. No oil storage tanks appear to be damaged.

No enemy aircraft were seen or encountered. Flak encountered at the PT was MAH for approximately 4 minutes. Eleven (11) aircraft of our formation were holed by flak, one of which - aircraft 834 - had an engine shot out. This aircraft left the formation after the target and returned over land via Rome, escorted by 12 P-38's.

Weather: There were a few scattered altocumulus clouds over the base. It was CAVU over the target.

Thirty-two (32) aircraft landed at 1535 hours. Aircraft 730 landed. at Corsica because of engine malfunction and returned to base at 1721 hours. There were no losses.

## MISSION NO. 54 - 3 AUGUST 1944

At 0656 hours, 38 B-24's (39 scheduled) took off to bomb the FRIEDRICHSCHAFEN MANZELL AIRCRAFT WORKS (PT) at Friedrichschafen, Germany. The 1st attack unit was led by Major Richard V. Griffin, 830th CO and the 2nd attack unit was led by Major David L. Sjodin, 831st CO. The Group assembled into Wing formation. Rendezvous with fighter escort approximately 50 P-51's - was effected at 0935 hours. The escort was last observed at 1218 hours. Eight aircraft returned early: WJ, RS, RM, YX, WI, WQ, RJ, BH.

Thirty (30) aircraft were over the target at 1113 hours and dropped 75 tons of 500 lb. GP bombs. Target was partially obscured by smoke from previous bombings. Crewmembers reported some strikes in target area with, some bombs falling to the southwest into the lake. Preliminary damage assessment shows: Target to be partially covered with smoke. Some bombs fell to the southwest into the water, with indications that some bombs fell into the smoke-covered area of the target.

About 40 to 50 enemy aircraft were in the target area and on return to the Adriatic coast. None of these fighters attacked the formation. However, they aggressively attacked the Group immediately behind. Flak encountered only over the target. Formation was in M to IIH flak for 3 to 4 minutes. Only two aircraft received slight flak damage. There were no casualties.

Weather: There were 1/10 altocumulus clouds at 16,000 feet over the target. There were 3/10 cumulus and a few towering cumulus clouds over the base on return.

Thirty (30) aircraft landed at base at 1414 hours. There were no losses.

## MISSION NO. 55 - 6 AUGUST 1944

At 0719 hours, 39 B-24's (40 scheduled) took off to bomb the TARASCON/RHONE RAILROAD BRIDGE (PT) at Tarascon, France. The 1st attack unit was led by Col. Arnold, GP CO and the 2nd attack unit was led by Major Dale S. Seeds, Asst. GP Ops Officer. Aircraft 791 failed to take off. The Group assembled into Wing formation. There was no fighter escort.

Thirty-eight (38) aircraft were over the target at 1153 hours and 37 aircraft dropped 90.5 tons of 500 lb. RDX bombs from 20,800 feet. Results: A direct hit on the bridge at the 2nd pier of the west end approach. Two direct hits on the 3rd span from the west end. Six strikes on tracks of the west end approach to the bridge. Three strikes on tracks at the east end approach to the bridge at track junction. Four strikes on tracks running to the south beyond the junction. The bomb Pattern straddled the bridge with bombs falling both short and over.

About 8 enemy aircraft were seen at landfall east of Toulon. Seven were unidentified single engine fighters and one was an Me 109, No enemy encountered.

Weather enroute there were a few patches of cumulus clouds at 7,000 feet.
It was clear over the target.

Thirty-five aircraft landed at 1,502 hours. Col. Arnold, in aircraft 127, left the formation after leaving the French coast on return with no. 4 engine feathered and landed at Rome. The aircraft has not returned to base. Major Cummings took over the lead and brought the formation back to base. Aircraft 416 landed at Pomigliano Field to hospitalize a crewmember who suffered an attack of appendicitis. Aircraft returned to base at 2028 hours. Aircraft 110 landed at Pomigliano Field because it was running short of fuel. Aircraft returned to base at 2051 hours.

## MISSION NO. 56 - 7 AUGUST 1944

At 0710 hours, 30 B-24's (32 scheduled) took off to bomb the BLECHHAMMER NORTH SYNTHETIC OIL PLANT in Germany. Red "X" failed to take off because the right landing gear collapsed while taxiing. Aircraft, yellow "P" failed to take off due to mechanical trouble. The 1st attack unit was led by Lt. Col. Robert E. Smith, GP Ops Officer and the 2nd attack unit was led by 1st Lt. Roy M. Smith, 828th Flt Cmdr. The Group assembled into Wing formation. Rendezvous with fighter escort - approximately 50 P-51's - was effected at 1015 hours. At 1212 hours, several P-38's joined, escorting the formation over the target and on withdrawal, leaving the formation, at 1430 hours.

Six aircraft returned. early: 503, 729, 791, 162, and 277. Aircraft 414 returned at 1303 hours with no. 4 engine out. This aircraft dropped its bombs on the Syambathely Airdrome in Hungary were approximately 25 JU 88's were dispersed, on the airdrome. No hits were observed.

Approximately 25 enemy aircraft - 15 Me 109's, 7 FW 190's, 1 Me 2l0, 1 Me 110 and 1. JU 88 were observed between the target and Lake Balaton. Five (5) enemy aircraft attacked the high box of the 2nd unit. Attacks were aggressive from 5 to 6 o'clock, low and singly.

Our gunners claimed three of these aircraft - 1 JU 88 and 2 Me 109's probably destroyed. Gunners reported that enemy aircraft attacking another Group used the following tactics. One aircraft would faint at a B-24 from behind and below, drawing fire from the ball turret. At the same time, another enemy aircraft would attack the aircraft from the opposite direction.

Two B-24's were observed shot down, by using these tactics.

Flak over the target was IAH - defenses increased. Formation was in flak for 5 minutes. Eleven (11) aircraft received minor damage from flak. Aircraft 721 received a direct hit from flak after releasing Its bombs on the target. It 'began to lag, later losing altitude in what appeared to be an attempt to join the formation at a lower altitude. Two enemy aircraft attacked this aircraft between Tranna and Lake Balaton, apparently firing rockets. Eight chutes opened at 1235 hours. When last seen, the pilot had the aircraft under control with enemy aircraft continuing the attack.

Twenty-four (24) aircraft were over the target at 1131 hours and dropped 59.75 tons of 500 lb. GP bombs from 22,600 feet. Preliminary damage assessment indicates that the target are was well covered with hits. Crews reported approximately 50 smoke generators around the target area. Due to wide dispersal of pots and to unfavorable wind conditions, smoke screen was ineffective.

Weather: There were a few patches of stratus clouds at 3,000 feet at assembly. It was clear in the vicinity of the target.

Twenty (20) aircraft landed at 1514 hours. Aircraft 394 and 508 landed at Vis and returned to base after refueling. Aircraft 344 returned late.

831st Bomb Sqdn Aircraft 721

P  1st Lt. Richard J. Erhardt
CP  2nd Lt. Waladyslaw Paterak
N  2nd Lt. Frederick Irving
B  2nd Lt. Morgan M. Browning
EWG  SSgt John D. Coughlin
RWG  Cpl. Charles H. Dahlgren
TG  Sgt. Carlio Dial
UG  PFC William R. Hering
NG  Sgt. John J. Battista
BTG SSgt Walter L. Dougherty

## MISSION NO. 57 - 9 AUGUST 1944

At 0754 hours, 29 B-24's (31 scheduled) took off to bomb the BUDAPEST/TOKOL AIRDROME, INSTALLATIONS (PT) in Hungary. The 1st attack unit was led by Major Nett, 828th CO and the 2nd attack unit was led by 1st Lt Donald R. Whiteman, 829th Flt Cmdr. The Group assembled into Wing formation. Rendezvous with, fighter escort - 20 P-38's - was effected at 1017 hours. The escort left the formation at 1230 hours. Fighter cover was close to the formation and very good. Four aircraft returned early prior to bombing: 791, 426, 724, and 394,

Twenty-five (25) aircraft were over the target at 1115 hours and dropped 62.5 tons of 500 lb. GP bombs from 22,600 feet. Photos revealed that the target was almost covered with smoke when the formation arrived. Bomb strikes began to fall just short of the target continuing beyond the runway, indicating that some bombs fell into the smoke obscured area, hitting the briefed point of impact. The southeast half of the runway was well covered by bomb strikes.

Three Me 109's were seen between the target area and did not attack the formation. Flak at the target was MAH, the formation was exposed to flak for 3 minutes. Nine aircraft were holed by flak, damage slight.

Weather: Target area was CAVU. It was clear over the base.

Twenty-five (25) aircraft landed at 1359 hours.

## MISSION NO. 58 - 10 AUGUST 1944

At 0710 hours, 28 B-24's, scheduled, took off to bomb the ATRO ROMANO OIL REFINERY (PT) at Ploesti, Rumania. The first attack unit was led by Major Thomas D. O'Brien, 829th CO and the 2nd attack unit was led by Major Dale S. Seeds, Asst. Gp Ops Officer. The Group rendezvous with the 460th BG. Rendezvous with fighter escort - 20 P-51's - was effected at 1045 hours. The escort left the formation at 1220 hours. Five aircraft returned prior to bombing: 504, 494, 460, 426, and 532.

Twenty-three (23) aircraft were over the target at 1106 hours and dropped 56.25 tons of 500 lb. RDX bombs. Results of the bombing: Two strikes on storage tank in the southeast corner of the target area and two near misses on the north storage tank in the group of four, south of the target.

Fifteen (15) Me 109's were seen in the target area but did not approach the formation. Sixteen (16) aircraft were holed by flak - damage minor. Flak encountered at the target was IAH. One (1) bomber is missing. Bombardier in no. 2 lead box, 1st unit, received a minor facial wound due to flak.

Weather: There was 3/10 cirrus clouds over the base and clear over the target.

Twenty-one (21) aircraft landed at 1432 hours. Aircraft 863 landed at a friendly airfield and the crew returned to base at 1930 hours.

829th Bomb Sqdn Crew Downed

P 2nd Lt. Francis C. Lozito
CP 2nd Lt. Elmer B. Hall
N 2nd Lt. Clifford J. Mandersheid

B 2nd Lt. William H. Donovan
EWG SSgt Kenneth E. Robinson
RWG TSgt Edgar J. Helms, Jr.
TG Sgt Richard M. Salasar
BTG SSgt Forrest L. Yeager
NG Sgt Thomas H. West
UG Sgt Billy R. Culver

## MISSION NO. 59 - 12 AUGUST 1944

At 0636 hours, 28 B-24's, scheduled, took off to bomb SETE GUN INSTALLATIONS (PT) in France. The 1st attack unit was led by Major John E. Atkinson, 831st CO and the 2nd attack unit was led by 1st Lt. Hugh L. Garnett, 830th Flt Cmdr. The Group rendezvous with the 460th BG. The 464th and 465th BG's were never sighted. Rendezvous with fighter escort - 15 P-51's - was effected at 1100 hours. The escort left the formation at 1200 hours. Aircraft 416 returned early, prior to bombing.

Twenty-seven (27) aircraft were over the target at 1112 hours and 26 aircraft dropped 65 tons of 500 lb. RDX bombs, between 17,000 and 17,400 feet. After reaching the IP, the Group broke off in four boxes. Results of bombing - not impressive. No enemy aircraft were seen and no flak encountered.

Weather: There were low stratus clouds over the base at take off time at 300 feet with tops at 1500 feet. It was clear over the target.

Twenty-seven (27) aircraft landed at 1500 hours.

## MISSION NO. 60 - 13 AUGUST 1944

At 0832 hours, 28 B-24's, scheduled, took off to bomb the SETE GUN INSTALLATIONS (PT) in France. The 1st attack unit was led by Major Griffin, 830th CO and the 2nd attack unit was led by Captain John M. Jones, 831st Ops Officer. The Group assembled into Wing formation. Rendezvous with fighter escort - 16 P-51's - was effected at 1300 hours. The escort left the formation at 1435 hours. Aircraft 791 and 725 returned prior to bombing.

Twenty-six (26) aircraft were over the target from 1311 to 1,317 hours. Twenty-five (25) aircraft dropped 62 tons of 500 lb. RDX bombs from between 16,550 to 17,600 feet. After reaching the IP, the Group broke off into four boxes. Bombing results were satisfactory. No enemy aircraft were seen. SIH flak was encountered over Sete for approximately one minute.

The weather clear over the base and the target.

Twenty-six (26) aircraft landed at base at 1659 hours.

## MISSION NO. 61 - 14 AUGUST 1944

At 1246 hours, 31 B-224's, scheduled, and took off to bomb the GUN POSITIONS in the St. Tropex area in France. The 1st attack unit was led by Lt Col Smith, Gp Ops Officer and the 2nd attack unit was led by Major Phillip E. Cummings, 828th CO. The Group assembled into wing formation. Four (4) P-51's joined the formation at 1557 hours and were not seen after target time. There were no early returns.

Thirty-one (31) aircraft were over the target at 1639 hours and 28 aircraft dropped 70 tons of 1000 lb. GP bombs from 17,000 feet. The results were not impressive. No enemy aircraft were seen and no flak was encountered.

Weather: There were 5/10 altocumulus clouds at 8,000 feet over the base and the target area was clear.

Thirty-one (31) aircraft landed at the base at 1940 hours.

## MISSION NO. 62 - 15 AUGUST 1944

At 0834 hours, 33 B-24's, scheduled, took off to bomb the PONT ST. ESPRIT HIGHWAY BRIDGE (PT) in France. The 1st attack unit was led by Col Arnold, Gp CO and the 2nd attack unit was led by Major Phillips E. Cummings, 829th Ops Officer. The Group assembled into Wing formation. Rendezvous with fighter escort - 25 P-51's - was effected at 1221 hours. The escort left the formation at 1323 hours. There were no early returns.

Thirty-three aircraft were over the target at 1257 hours and 31 aircraft dropped 77 tons of 1000 lb. GP bombs from 15,000 feet. The bridge was destroyed. No enemy aircraft were seen and no flak was encountered.

Weather: It was clear over the base and there were 2/10 cumulus clouds over the target area.

Thirty-three aircraft landed at base at 1702 hours.

## MISSION NO. 63 - 16 AUGUST 1944

At 0726 hours, 28 B-24's took off to bomb the OBER/RADERRACK CHEMICAL WORKS (PT) at Friedrichshafen, Germany. The 1st attack unit was led by Lt Col. Herblin, Deputy Gp CO and the 2nd attack unit was led by 1st Lt Hugh L. Garnett, 830thFlt Cmdr. The Group assembled into wing formation. Twenty-five (25) P-38's joined the formation at 1000 hours, furnishing penetration and target coverage. Several P-51's joined the formation at 1300 hours, providing withdrawal cover. Five aircraft returned prior to bombing: 729, 724, 791, 394 and 426. No enemy aircraft were seen. The flak over the target was MIH, lasting for 4 minutes. Crews reported seeing peculiar flak bursts, or flares over the PT. These flares were yellow in color, bursting above the formation and floating slowly down, burning with a bright yellow flame and leaving a white smoke trail behind. The rate of descent, suggested that the flares were suspended by parachutes.

Twenty-three (23) aircraft were over the target at 1128 hours and 18 aircraft dropped 43.5 tons of 500 lb. RDX bombs. The results were generally fair. Some bombs fell over several thousand yards short of the target. However, some strikes were in the immediate target area.

Weather: It was clear over the base, and clear over the target except for high cirrus clouds.

Twenty-three aircraft landed at base at 1452 hours.

## MISSION NO. 64 - 18 AUGUST 1944

At 0648 hours, 28 B-24's. scheduled, took off to bomb OIL INSTALLATIONS at Ploesti, Rumania. The PT was the ROMANA AMERICANO OIL REFINERY. The 1st attack unit was led by Major Reeve, Asst. Gp Ops Officer and the 2nd attack unit was led by 1st Lt. William C. Lawrence, 831st Flt Cmdr. The Group assembled into Wing formation. Several P-51's joined the formation at 0945 hours and left the formation at 1245 hours.

Some difficulty was encountered enroute, because the lead Group was flying too slow. The 485th had to continuously fly "esses" in order to stay in formation. Major Reeve called the Wing Leader, reporting this condition and it was corrected. Three (3) aircraft returned prior to bombing: Aircraft 139, piloted by Major Reeve; Deputy Leader, 1st Lt. Clark R. Miller, 829th Flight Cmdr took over the lead and led the formation over the target. Aircraft 728 and 394, were the other two aircraft to return early. These three aircraft were credited with sorties, inasmuch as each received minor flak damage over new territory.

Twenty-five (25) aircraft were over the target at 1035 hours and 24 aircraft dropped 48 tons of 500 lb. GP bombs from 25,100 feet by PFF. Results: Several bombs hit in the target area, most of which landed in the finished product storage section. Three Me 109's were seen near Ploesti. However, none attacked the formation. Flak over the target was IAH for 4 minutes. MAH flak was encountered from railroad guns on the railroad east of Svode. Twelve (12) received minor flak damage. There were no casualties.

Weather: It was clear over the base with high-scattered clouds over the target area.

Twenty-five (25) aircraft landed at base at 1351 hours.

* "fly esses" - weaving back and forth to slow down

## MISSION NO. 65 - 21 AUGUST 1944

At 0703 hours, 32 B-24's, scheduled, and took off to bomb the NIS AIRDROME (PT) IN Yugoslavia. The 1st attack unit was led my Major Nett, 828th CO and the 2nd attack unit was led by 1st Lt. Lawrence D. Richins, 830th Pilot. The Group assembled into Wing formation from 0955 to 1045 hours. There were no early returns.

Thirty-two (32) aircraft were over the target at 1001 hours and dropped 68.52 tons of Fragmentation Bombs - 36 Frag Clusters - from 20,500 feet. The results were excellent. The bomb pattern started just short of the ramp, continuing across aiming point and the hangar installations. Some bombs fell beyond the target are near the marshalling yard. No enemy aircraft were seen. Flak at the target was SIH and five (5) were holed by flak - damage slight.

Weather: It was clear over the base and the target area.

Thirty-two (32) aircraft landed at 1200 hours.

## MISSION NO. 66 - 22 AUGUST 1944

At 0611 hours, 28 B-24's, scheduled, and took off to bomb the KORENEUBURG OIL STORAGE (PT) at Vienna, Austria. The 1st attack unit was led by Major Griffin, 830th CO and the 2nd attack unit was led by Major Ralph E. Monroe, 828th Ops Officer. Col. Arnold, Gp CO, flew in the lead aircraft. The Group rendezvous with the 460th BG. Twenty (20) P-51's and P-38's escorted the formation from 0930 to 1130 hours. Three aircraft returned prior to bombing: 474, 344 and 277.

Twenty-five (25) aircraft were over the target at 1017 hours and 23 aircraft dropped 57.5 tons of 500 lb. GP bombs from 23,000 feet. The results were poor - with no bomb strikes in the target area. No enemy aircraft were seen. The flak over the target was M-IAH Fourteen (14) aircraft were holed by flak - damage minor. One aircraft dropped behind the formation and is missing. Aircraft 842 was last seen to drop out of formation just after the rally. It continued to follow the formation and was under control, but it dropped back gradually until it was out of sight of the formation. Aircraft 344 and 921 made emergency landings at Vis. The pilots of these crews were evacuated to base with the aircraft remaining at Vis for repairs.

Weather: There were a few patches of cumulus clouds over the base and clear over the target area.

Twenty-two (22) aircraft landed at base at 1306 hours.

830th Bomb Sqdn Aircraft 842

P   1st Lt. Edwin C. Tuttle
CP  2nd Lt. Lawrence D. McGilvary
N   2nd Lt. Frank E. Gallager
B   2nd Lt. Oscar Rutstein
RWG TSgt Garnie Martin
UG  SSgt Frank J. Brockel
WG  TSgt Donald B. Landrum
WG  SSgt Leon W. Hoadley
BTG SSgt Clifford Brown
TTG SSgt Hershel L. Hasenfuss

## MISSION NO. 67 - 23 AUGUST 1944

At 0826 hours, 28 B-24's, scheduled, took off to bomb the MARKERSDORF AIRDROME in Austria. The 1st attack unit was led by Major O'Brien, 829th CO and the 2nd attack unit was led by 1st Lt. Roger A. Nichols, 828th Flt Cmdr. Both leaders were forced to turn back prior to bombing and the lead positions were taken over by 1st Lt Clark R. Miller, 829th Flt Cmdr and Captain Clyde E. Cribley, 828th Flt Cmdr. The Group assembled into Wing formation. Twenty (20) P-51's escorted the formation from 1130 to 1400 hours. Aircraft 138, 818, and 498 returned early.

Twenty-five (25) aircraft were over the target at 1236 hours. Because the target was cloud obscured, only 8 aircraft dropped 17.25 tons of 500 lb. GF bombs. Sixteen (16) aircraft returned 160 bombs to base. Results of the bombing were unobserved. No flak was encountered or no enemy fighters were encountered.

Eight (8) Me 109's were seen in the target area and were engaged by the P-51's before they could attack the formation. Right over the target, at an altitude of 23,400 feet, a sliver colored aircraft, believed to be an Me 163, having a short-not too stubby and very sleek fuselage width swept back wings, which left a "smoke ring" contrail, made a pass over the high box of the second unit from 5 o'clock, then went up at an 85 degree angle of attack to attack five P-51's, flying escort for the formation. The last P-51 in the formation broke out and looped over and under and attacked and shot down the Me 163 while it was in its climb to attack the other four aircraft. When hit, the Me 163 broke into three pieces - the wings, which folded back over the fuselage, and the fuselage. The estimated speed of thee Me 163 was at least 400 mph.

Weather: There were 4-6/10s cumulus and altocumulus clouds over the target with visibility of 5 to 7 miles. There were a few cumulus clouds over the base.

Twenty-five (25) aircraft landed at base at 1552 hours.

## MISSION NO. 68 - 24 AUGUST 1944

At 0751 hours, 28 B-24's, scheduled, took off to bomb the PARDUBICE OIL REFINERY (PT) in Czechoslovakia. The 1st attack unit was led by Major Atkinson, 831st CO and the 2nd attack unit was led by 1st Lt. Donald Whiteman, 829th Pilot. The Group assembled into Wing formation. Thirty (30) P-38's escorted the formation from 0826 to 1355 hours. Aircraft 728 returned early.

Twenty-seven (27) aircraft were over the target at 1219 hours and dropped 67.5 tons of 500 lb. RDX bombs. Results: The bomb pattern consisted of forty to fifty strikes located at the extreme northwest corner of the target area. The target area was smoke covered.

About 30 to 40 enemy aircraft were seen between the rally point and Linz, Austria, all of which were Me 109's and FW 190's. Thirty (30) of these attacked the wing formation and the remainder were engaged by the fighter escort. The first attack was the mass formation, on the entire Wing formation from 4 to 6 o'clock, some level and some high. The enemy fighters then broke up and attacked single and in pairs. Attacks lasted from 1221 hours to approximately 1300 hours. All attacks were aggressive. Just before the attack of enemy aircraft came on our formation, the P-38's and P-51's were engaged by a number of enemy aircraft at approximately 25,000 feet. Our formation claims 5 FW 190's and 2 Me 109's destroyed; 1 FW 190 and 3 Me109 probables; 1 F 190 and 2 Me 109's damaged. Some FW 190's were silver colored, others were battleship gray fuselage and a dirty white belly, and some were OD color. No flak was encountered. One aircraft, 501, is missing. This aircraft was damaged by enemy fighters in the first pass on the formation at 1220 hours. No. 1 and 2 engines were on fire. The fire on no.1 engine was put out and the aircraft dropped behind the formation. At 1234 hours, the crew was seen to bail out. One crew reported 10 chutes and another crew reported 7 chutes seen to open. There was one casualty - the Bombardier in no. 5 aircraft, in the high box of the 2nd unit - received a minor scalp wound.

Weather: There were 2/10 decay cumulus clouds over the base. The target area was smoke covered.

Twenty-six (26) aircraft landed at base at 1530 hours.

831st Bomb Sqdn Aircraft 501

| | | | |
|---|---|---|---|
| P | 1st Lt James E. Mulligan | ENG | SSgt Robert R. Rector |
| CP | 2nd Lt Robert A. Seitz | TG | Sgt Leonard L. Little |
| N | 2nd Lt Sammuel P. Giaimo | UG | Sgt Slavko (NMI) Nenadich |
| B | 2nd Lt Francis j. Nardi | BTG | Sgt Leo J. Gagne |
| RO | SSgt John J. Godfrey | NG | Cpl. Donald D. Evjen |

## MISSION NO. 69 - 25 AUGUST 1944

At 0648 hours, 28 B-24's, scheduled, took off to bomb the PROSTEJOV AIRDROME INSTALLATIONS (PT) in Czechoslovakia. The 1st attack unit was led by Major O'Brien, 829th CO and the 2nd attack unit was led by Captain John M. Jones, 831st Ops Officer. The Group assembled into Wing Formation. Thirty (30) P-51's and P-38's escorted the formation from 0952 to 1351 hours. Aircraft 422 and 526 returned prior to bombing.

Twenty-five (25) aircraft were over the target at 1131 hours and 23 aircraft dropped 57.5 tons of 500 lb. GP bombs. Results: The target was completely smoke covered. Forty to sixty bomb strikes were in the landing area south of the target and 15 to 20 strikes were definitely in the target area with possibly more in the smoke-covered area. About 5 enemy aircraft were seen in the target area: 3 Me 109's and 2 FW 190's. No enemy aircraft were encountered and no flak was encountered.

Weather: There were 3/10 cumulus clouds over the base and 2/10 cloud coverage over the target area.

Aircraft 494 landed at Vis. Twenty-five (25) aircraft landed at base at 1442 hours.

## MISSION NO. 70 - 27 AUGUST 1944

At 0730 hours, 28 B-241s, scheduled, took off to bomb the BLECHHAMMER SOUTH OIL REFLNERY (PT) in Germany. The 1st attack unit was led by Col Arnold, Gp CO and the 2nd attack unit was led by Captain Francis P. Tunstall, 830th Flt Cmdr. The Group assembled into Wing formation. Seven (7) P-51's escorted the formation from 1120 to 1310 hours. Four aircraft returned prior to bombing: 638, 486, 157, and 474.

Twenty-four (24) aircraft were over the target at 1214 hours and dropped 59.75 tons of 500 lb. GP bombs from 22,600 feet. Results were excellent with all bombs landing in the target area. Flak over the target was IAH, lasting for 5 minutes. The lead aircraft of the formation, piloted by Col Arnold, received several flak hits in the nose section and in the front bomb bay, a few seconds before bombs away. A good course had been set up and when the bombardier of the deputy lead aircraft saw that the leader was going to be late getting his bombs away, released approximately 3 seconds ahead of the leader. As soon as bombs were away, the leader put his aircraft into a steep dive to the right, losing 2,000 feet of altitude before he leveled off. The formation stuck with the leader, thinking that he was rallying. They did not realize that the aircraft was in extreme difficulty until crew members began to leave the aircraft. After all crew members bailed out, the aircraft went down and crashed at 1215 hours. Smoke generators were in operation at the target.

On return, aircraft 517, while over the mountains in Yugoslavia, had all four engines temporally cut out at the same time. The aircraft lost 3,000 feet before the engines could be started, causing much anxiety to the crew members. Aircraft 127 was lost to flak. Aircraft 162 landed at Vis, after failing to keep up with the formation. Fourteen (14) aircraft were holed by flak: Twelve (12) aircraft received minor damage and 2 aircraft received major damage.

Weather: There were 3/10 cumulus clouds over the base and 2/10 cloud coverage over the target area.

Twenty-two (22) aircraft landed at base at 1555 hours.
828th Bomb Sqdn Aircraft 127

| | | | |
|---|---|---|---|
| P | Col Walter E. Arnold, Jr. | ENG | TSgt Ben F. Cook |
| CP | Lt Col Robert A. Smith | RO | TSgt William R. Killian |
| B | Captain Kenneth N. Gillespie | AG | SSgt Theodore A. Brown |
| N | Captain Jay (nmi) Jaynes | AE | SSgt Shirley W. Hancock |
| NN | Captain John W. Coffin | AE | SSgt Quentin E. Meyer |
| PN | 1st Lt. Warren D. Sortomme | NG | SSgt Tony (nmi) Annie |

## MISSION NO. 71 - 28 AUGUST 1944

Twenty-eight (28) B-241s, scheduled, took off to bomb the SZONY OIL REFINERY (PT) in Hungary. The 1st attack unit was led by Major Monroe, 828th Ops Officer and the 2nd attack unit was led by Major Dale S. Seeds, Asst. Gp Ops Officer. The Wing assembled into Wing formation. Twenty-five (25) P-38's escorted the formation from 1000 to 1150 hours. Aircraft 730 returned prior to bombing.

Twenty-six (26) aircraft were over the target at 1043 hours and dropped 65 tons of 500 lb. GP bombs from 22,400 feet. Results were very good with all bombs falling in the target area. Aircraft 729 turned back at 0925 hours and landed safely at Vis.

No enemy aircraft were seen. SIH to SAH flak was encountered over the target for about one minute. Three aircraft received minor flak damage. An unexploded 88mm shell passed through the nose compartment of aircraft 504. The Bombardier, 2nd Lt John M. Veal, received a wound in the leg from this shell. He was landed at Foggia in serious condition and died a few minutes later.

Weather: CAVU over the base and target.

The formation was over the base at 1317 hours. The seventh aircraft to land, landed with the rose wheel unlocked and blocked the runway for 45 minutes. All aircraft landed by 1403 hours. Aircraft 504 landed late.

## MISSION NO. 72 - 29 AUGUST 1944

At 0716 hours, 28 B-24's, scheduled, took off to attack the MORAVSKA/OSTRAVA MARSHIALLILNG YARD in Czechoslovakia. The 1st attack unit was led by Lt Col Herblin, Deputy Gp CO and the 2nd attack unit was led by Major Nett, 828th CO. The Group assembled into Wing Formation. Twenty (20) P-51's escorted the formation from 1110 hours and several P-38's joined after the target at 1120 hours. All escort left the formation at 1312 hours. Aircraft 394 and 755 returned prior to bombing.

Twenty-six (26) aircraft were over the target at 1118 hours. The formation did not drop their bombs because of a malfunction of equipment in the lead aircraft. The deputy lead also had a malfunction and there were no other PFF aircraft in the formation. Nineteen (19) aircraft returned their bombs to the base and seven (7) aircraft jettisoned their bomb load. Four unidentified aircraft were seen in the Budapest area. No enemy aircraft were encountered. Flak was scant to moderate, inaccurate, heavy. There was no damage.

Weather: It was CAVU over the base and there was 8-9/10 cloud coverage over the target area.

Twenty-six (26) aircraft landed at base at 1455 hours.

## UNIT HISTORY, 485TH BG (H) 1 SEPTEMBER 1944 TO 30 SEPTEMBER 1944

The 485th BG flew 14 operational missions during the month, making a grand total of 86 missions as of 30 September 1944. 914.73 tons of bombs were dropped on numerous targets, with bombing results being very good. According to 15th AF statistical publication, "Bombing Accuracy", of 25 September 1944, the Group was third highest in the 15th AF.

Six (6) aircraft and their crews were lost in action.

Seven (7) officers and 10 enlisted men were evacuated from a POW camp - Main Camp Choumen, in Bulgaria. Major Walter A. Smith, Jr., who bailed out over Yugoslavia on 23 June 1944 was officer in charge of the prisoners, under the Bulgarian Commandant. Major Smith organized a camp staff and daily routines. The accommodations were very inadequate for the men there, and he tried constantly to improve them. Conditions, in general, improved greatly as the war improved for the Allies. On August 2nd, he organized a news supply system by bribing the guards and the camp was fairly well supplied with information. On 8th September, Major Smith was piloted by a Bulgarian Captain in a German Stork aircraft to Sofia. At Sofia, he contacted the Chief of Staff of the Bulgarian Army and the Chief of Operations, and with them, he arranged by telegraph the movement of POW's from Choumen, going by train into Turkey.

Three (3) officers and 3 enlisted men were also evacuated from a POW Camp in Rumania. All personnel from these camps were immediately sent back to the States.

The Group did not encounter any enemy aircraft during the month. The following claims were submitted by men returning from POW camps. These enemy aircraft were shot down before the men bailed out over enemy territory. 6 e/a destroyed.

On 27 September 1944, the Group celebrated its 1st anniversary. The event was highlighted by the visit of numerous distinguished guests, among them being:
B/G George R. Acheson, C/G 55th BW; Col. William L. Lee, CO 49th BW; Col. J. M. Price, CO, 460th BG and Col. A. L. Schroeder, Co 464th BG.

On 15 September, Group Special Services undertook to construct the Group Theatre and Gymnasium.

AWARDS:
    Silver Star - 13 Officers and. Enlisted Men
    DFC - 323 Officers and Enlisted Men
    Bronze Star 1 Officer

Twenty-six (26) new crews arrived at the base.

**MISSION 74, 14 SEPTEMBER 1944, TARGET: ADIGE/ORA RAILROAD BRIDGE (across the Adige River in northern Italy). Bombs: 1000 lb RDX, bombing altitude 21,000 feet.**

Ora is located on the Adige River (the white strip in the photo) in the Brenner Valley, which was one of the main routes through the Alps used by the Germans to supply their troops on the Italian Front. Sometimes the 1000 lb bombs had delayed fuses to delay the rebuilding of the bridges. Leaflets were also dropped to warn the civilians.

**MISSION 76, 6 SEPTEMBER 1944. TARGET: NYIREGYHAZA MARSHALLING YARDS IN HUNGARY. BOMBS: 500 LB. RDX FROM 16,500 FEET.**

On this mission our 55th Bomb Wing participated with four other wings (21 bomb groups) in a coordinated attack on communications in Hungary and northern Yugoslavia, to support the Russian forces in securing a favorable position for the destruction of German and Hungarian forces in eastern Hungary, and to create a favorable situation for their drive westward on Budapest. The immediate tactical objective of the attack on Niregyhaza was the destruction of rolling stock and supplies contained therein, to create a temporary transportation block at the focal point, and to deny the enemy the use of supplies at the marshalling yards on their way to the front.

One might ask why the bombs didn't all fall on the target. The technology and equipment that was used during WWII would be considered to be antiquity by today's standards. A number of bombs were used to assure the destruction of the target. The cost of the bombs was insignificant compared to the cost of delivering them. In addition, the shock wave produced by the impact of the bombs falling close to the target also caused damage to the surrounding buildings.

**MISSION NO. 73 - 1 SEPTEMBER 1944**

At 0751 hours, 28 B-24's, scheduled, took off to bomb the SZAJOL MARSHALLINIG YARD (PT) in Hungary. The 1st attack unit was led by Major Reeve, Gp Ops Officer and the 2nd attack unit was led by Major Cummings, 829th CO. The Group assembled into Wing formation. Rendezvous with fighter escort - 20 P-51's - was effected at 1031 hours. Aircraft 885 returned to base prior to bombing.

Twenty-seven (27) aircraft were over the target at 1103 hours and 25 aircraft dropped 62.5 tons of 500 lb. GP bombs by visual means. Results: The marshalling yard was well covered with bomb strikes. No enemy aircraft were seen. SIH flak was encountered for 1 minute on the rally to the left from the target. None of the aircraft were damaged by flak.

The weather was CAVU over the base and the target area.

Twenty-seven (27) aircraft landed at 1342 hours.

**MISSION NO. 74 - 4 SEPTEMBER 1944**

At 0958 hours, 28 B-24's, scheduled, took off to bomb the ADIGE/ORA RAILROAD BRIDGE (PT) in northern Italy. The 1st attack unit was led by Lt. Col. John P. Tomhave, Gp CO and the 2nd attack unit was led by Lt. William C. Lawrence, 831st. Flt Cmdr. The Group assembled into Wing formation. Thirty (30) to 40 P-51's escorted the formation from 1225 to 1403 hours. There were no early returns.

Twenty-one (21) aircraft were over the target at 1325 hours and dropped 52.5 tons of 1000 lb. RDX bombs from 21,000 feet. The results were very good. Both approaches were hit, a direct hit on the center span and a direct hit on the adjacent highway bridge. The lead box of the 2nd unit over shot its position on the bomb run and bomb another highway bridge, 3 miles south of the primary target. No enemy aircraft were seen and no flak was encountered.

Weather: There was a 6/10 cloud coverage over the base. It was clear over the target area.

Twenty-eight (28) aircraft landed at base at 1608 hours.

**MISSION NO. 75 - 5 SEPTEMBER 1944**

At 0700 hours, 31 B-24's, scheduled, took off to bomb the SZOB RAILROAD BRIDGE (PT) in Hungary. The 1st attack unit was led by Major Atkinson, 831st CO and the 2nd attack unit was led by Captain Hugh L. Garnett, 830th Flt Cmdr. The Group assembled into Wing formation. Thirty (30) to 40 P-51's escorted the formation from 1000 to 1240 hours. There were no early returns. Thirty-one (31) aircraft were over the target at 1051 hours and dropped 73.5 tons of 1000 lb. RDX bombs. The target was smoke obscured from bombing by the lead Groups and it was impossible to accurately determine the damage inflicted. Both approaches received several hits. No enemy aircraft were seen and no flak was encountered over the target. Near the IP, SIH to SAH flak was encountered and two aircraft received minor flak damage. There were no casualties.

The weather was clear over the base and target area.

On return, the 2nd aircraft to land, dropped two bombs on the runway, which failed to release over the target. The bombs were removed from the runway and 31 aircraft were landed by 1350 hours.

## MISSION NO. 76 - 6 SEPTEMBER 1944

At 0716 hours, 28 B-24's, scheduled, took off to bomb the NYIREGYHAZA MARSHALLING YARD, (PT) in Hungary. The 1st attack unit was led by Lt. Col. Nett, 828th CO and the 2nd attack unit was led by Captain George N. Thompkins, 830th Flt Cmdr. The Group assembled into Wing formation. Thirty (30) P-51's escorted the formation from 1000 to 1330 hours. Aircraft 422 and 277 returned prior to bombing.

Twenty-six (26) aircraft were over the target at 1114.5 hours and dropped 65 tons of 500 lb. RDX Bombs from 16,500 feet. The target was partially obscured by smoke from the bombing of the lead Groups, which prevented accurate damage assessment though it is apparent that much damage was inflicted. The aiming point was hit. No enemy aircraft were seen and no flak was encountered.

Weather: There were scattered cirrus clouds over the base and it was clear the in the target area.

Twenty-six (26) aircraft landed at base at 1440 hours.

## MISSION NO. 77 - 10 SEPTEMBER 1944

At 0700 hours, 28 B-24's took off to bomb a primary target, but due to adverse weather conditions, the 1st alternate target, the SOUTHEAST INDUSTRIAL AREA of Vienna, Austria, was bombed. The 1st attack unit was led by Major O'Brien, 829th CO and the 2nd attack unit was led by Major Reeve, Gp Ops Officer. The Group assembled into Wing formation. Twenty-eight (28) P-38's escorted the formation from 0901 hours to 1200 hours. Aircraft 899, 438 and 834 returned prior to bombing.

Twenty-five (25) aircraft were over the target at 1041 hours and dropped 62.5 tons of 500 lb. RDX bombs from 21,000 feet by PFF. Results were poor with military damage consisting of two strikes on the highway bridge approach over the canal.

Four (4) Me 109's were seen in the target area and were driven off by P-38's before they could attack the formation. IAH flak was encountered over the target for 7 - 10 minutes. Aircraft 532, after experiencing difficulty maintaining its formation position enroute, was apparently hit by flak after releasing its bombs over the target. It straggled on return and was last seen over the Adriatic Sea at 1240 hours. At that point, several members were seen to bail out. One crew reported hearing Air-Sea Rescue acknowledge a request for help from this aircraft. The aircraft was not seen to crash or to ditch. Aircraft 508 landed at Pantanella to discharge a wounded crew member and later returned to base. Cpl. C. C. Hard, Eng., received a wound from flak in his right forearm. Aircraft 276 landed at the same base with wounded crew members: TSgt F. Gross, RO - flak wound in right shoulder; Sgt J. M. Griffin, NG, abrasion of cornea of the left eye; and TSgt J. M. Leturno, EWG abrasion of the left forearm. Ten aircraft received minor flak damage. All aircraft returned to base, with 22 aircraft landing at 1355 hours.

830th Bomb Sqdn Aircraft 29532

| | | | |
|---|---|---|---|
| P | 2nd Lt. Arthur W. Rasco | WG | Cpl. Edward J. Kelly |
| CP | 2nd Lt. Robert W. Walkotte | TG | SSgt Raymond S. Lonergan |
| B | 2nd Lt. Joseph Stewart | BG | Cpl. Harold J. Kempffer |
| N | 2nd Lt. Max L. Childers | NG | Cpl. Raymond J. Buster |
| UG | TSgt Ernest W. Birch | WG | Cpl. Bloyce F. Jordan |

## MISSION NO. 78 - 12 SEPTEMBER 1944

At 0817 hours, 28 B-24's and 4 spare aircraft took off to bomb the WASSERBERG JET PROPELLER AIRCRAFT FACTORY (PT) in Germany. The 1st attack unit was led by Col. Tomhave, Op CO and the 2nd attack unit was led by Major Dale Seeds, Asst. Gp Ops Officer. The Group assembled into Wing formation. Twelve (12) P-38's escorted the formation from 1123 - later joined by several P-38's - until 1408 hours. The four spare aircraft returned to base and aircraft 507 returned to base prior to bombing.

Twenty-seven (27) aircraft were over the target at 1308 hours and 26 aircraft dropped 65 tons of 1000 lb. GP bombs from 21,000 feet. Direct hits were obtained on the factory buildings from a good bomb pattern. No enemy aircraft were seen and no flak was encountered over the target. Near Balzona, IAH flak was encountered for three minutes, coming from along the double railroad track connecting Balzano and the Brenner Pass.

Aircraft 819 had its control cables cut in the left side of the bomb bay by flak. Aircraft 755 had no. 1 engine shot out. Also, aircraft 138 lost an engine to flak. Aircraft 78384 on return ran out of gas and crashed about 20 miles south of the base. Seven (7) crewmembers escaped without injury. Two men were pinned in the aircraft. Cpl. Alexander E. Struich was killed and Cpl. Ralph J. Cox was seriously injured. Aircraft 330 landed at Vis for fuel and returned to base later. Aircraft 572 landed at Foggia for fuel and returned to base later. Aircraft 564 landed at Amadonna for fuel and returned to base later.

Weather: There were 2 10s cumulus clouds over the base and it was clear over the target area.

Twenty-three (23) aircraft landed at 1620 hours.

## MISSION NO. 79 - 13 SEPTEMBER 1944

At 0635 hours, 28 B-24's took off to bomb the OSWIECIM SYNTHETIC OIL and RUBBER PLANT (PT) in Poland. The 1st attack unit was led by Lt. Col. Herblin, Deputy Gp CO and the 2nd attack unit was led by Captain Vern E. Bryson, 829th Flt Cmdr. Also, 2 spare aircraft took off. The Group assembled into Wing formation. Thirty (30) P-38's joined the formation at 0930 hours and at 1025 hours, 30 P-51's joined the formation. The escort left the formation at 1253 hours. The two spare aircraft returned to base plus six aircraft returned prior to bombing: 486, 536, 495, 416, 819, and 474.

Twenty-two (22) aircraft were over the target at 1121 hours and 20 aircraft dropped 50 tons of 500 lb. RX bombs from 24,000 feet. The bomb run was begun with the aid of PFF as the target was obscure by smoke and clouds. The bombing results were poor with the main bomb pattern 3.5 Miles west of the briefed MPI. There were three strikes in the Marshalling yard, damaging at least 5 wagons. Also, there were 10 strikes in probable barracks area, damaging several buildings. No enemy aircraft were seen. IAH flak was encountered over the target for three minutes.

Aircraft 51139 received a direct hit between no. 3 and no. 4 engines about thirty seconds before bombs away. Fire started, after bombs were away and the aircraft began a slow spiral descent under control at Rudno, the rally point. Five chutes were seen to open before the aircraft disappeared at 1126 hours.

Aircraft 52601, evidently hit by flak over the target, feathered an engine and dropped out of formation. Ten (10) chutes from this aircraft were reported to have opened over Yugoslavia, at 1220 hours. The aircraft was not seen to have crashed.

Eleven (11) aircraft received minor flak damage.

2nd Lt. Wayne A. Wolf, co-pilot of aircraft 277 received a minor flak wound in the neck.

Weather: There was a 4/10s cloud coverage over the base and 5-6/10 coverage over the target area. Twenty (20) aircraft landed at 1346 hours.

831st Bomb Sqdn Aircraft 51139
P    Capt. William C. Lawrence
CP   1st Lt. Mathew W. Hall
N    2nd Lt. Daniel N. Blodgett
B    2nd T Frank J. Pratt
NN   1st Lt. George V. Winter
MICKEY OP  2nd Lt. Irving P. Canin
RO   TSgt William R. Eggers
UG   SSgt Vernon O. Christensen
WG   TSgt Everett L. McDonald
WG   SSgt Louis L. Kaplan
TG   SSgt Arthur E. Nitsche

831st Bomb Sqdn Aircraft 52601
P    2nd Lt. Jack Carter
CP   2nd Lt. William G. Barbour
N    2nd Lt. Kenneth W. Savee
B    2nd Lt. Joseph P. Lee
ENG  Sgt Artie Poulos
RO   Cpl. William R. Schellhorn
UG   Cpl. Reynold Solin
NG   SSgt Joseph Leutgenau
BG   Cpl. James H. Favre
TG   Cpl. John W. Sorrell, Jr.

## MISSION NO. 80 - 17 SEPTEMBER 1944

At 0810 hours, 28 B-24's, scheduled, and 3 spare aircraft took off to bomb the MAGYAR OIL REFINERY in Budapest, Hungary (PT). The 1st attack unit was led by Major Reeve, Gp Ops Officer and the 2nd attack unit was led by Captain Edward Stauverman, Jr., 831st Flt Cmdr. The Group assembled into Wing formation. Forty (40) P-38's escorted the formation from 1108 to 1300 hours. The spare aircraft returned to base plus six aircraft returned prior to bombing: 834, 725, 728, 438, 829, and 727.

Twenty-two (22) aircraft were over the target at 1156 hours and dropped 55 tons of 500 lb. GP bombs. The target was obscured by smoke from previous bombing. Several strings of bombs started hitting at the west-end of the marshalling yard adjacent to the refinery and extended into the target area. No strikes were noted outside of the target area. No enemy aircraft were seen. M-IH flak was encountered for 5 - 7 minutes over the target. Six aircraft received slight flak damage.

Weather: There were clouds at 1,000 feet over the base and a few clouds over the target area.

Twenty-two (22) aircraft landed at base at 1440 hours.

### MISSION NO. 81 - 18 SEPTEMBER 1944

At 0645 hours, 28 B-24's and 3 spare aircraft took off to bomb the BUDAPEST NORTH RAILROAD BRIDGE (PT) in Hungary. The 1st attack unit was led by Col. Nett, 828th CO and the 2nd attack unit was led by Captain Francis X. Dalton, 829th Flt, Cmdr. The Group assembled into Wing formation. Thirty to forty P-38's and P-51's escorted the formation from 0905 to 1245 hours. The 3 spare aircraft and aircraft 827, 068, and 299 returned to base.

Twenty-five (25) aircraft were over the target at 1127 hours and 13 aircraft dropped 32 tons of 1,000 lb. RDX bombs from 22,000 feet. Twelve (12) aircraft dropped 24 tons of 2000 lb. RDX bombs from 22,000 feet. Results: At least one direct hit near east end of the bridge and 5 - 7 near misses with probable damage to two piers. West approach to the bridge was cut by two direct hits.

On return 2 - 4 Me 109's, green in color, were seen at the Crehote and were heading in the opposite direction at 1300 hours. IAH flak was encountered for 5 minutes. Eight (8) aircraft received slight flak damage. Aircraft 50885 landed at Vis as gauges indicated shortage of fuel. The gauges were found to be malfunctioning and the aircraft returned later to base.

Weather: The rain showers over the base during take off with the base of the clouds at 1500 feet. The weather was clear over the target area.

Twenty-four aircraft landed at 1415 hours.

### MISSION NO. 82 - 20 SEPTEMBER 1944

At 0745 hours, 28 B-24's and 3 spare aircraft took off to bomb the HATVAN NORTH MARSHALLING YARD (PT) in Hungary. The 1st attack unit was led by Major Atkinson, 831st CO and the 2nd attack unit was led by 1st Lt. Eugene B. Lenfest, 830th Flt Cmdr. The Group assembled into Wing formation. Thirty to thirty-five P-38's joined the formation at 1150 hours and later several P-38's and P-51's joined and departed from the formation at 1300 hours. The 3 spare aircraft returned to base and there were no early returns.

Thirty-one (31) aircraft were over the target at 1151 hours and 15 aircraft dropped 37.5 tons of 500 lb. RDX bombs from 13,500 feet. Fifteen (15) aircraft dropped 37.5 tons of 500 lb. cluster incendiary bombs from the same altitude. Results in general were very good. Several strike in the marshalling yard. No enemy aircraft were seen and no flak was encountered.

Weather: There were 8/10s cirrus clouds over the base and 3/10s cirrus clouds over the target area.

Thirty (30) aircraft landed at 1505 hours. The 31st aircraft circled the field until all other aircraft landed. This aircraft had a shortage of hydraulic fluid. Landing was accomplished without mishap.

## MISSION NO. 83 - 21 SEPTEMBER 1944

At 0745 hours, 28 aircraft and 4 spare aircraft took off to bomb the DEHRECEN MARSHALLING YARD, (PT). The 1st attack unit was led by Lt. Col. Richard V. Griffin, 830th CO and the 2nd attack unit was led by Captain Vooney W. Wiggins, 831st Flt Cmdr. The other Groups of the Wing were recalled due to weather conditions, but permission was granted to the 485th Gp Leader to bomb the 1st Alternate Target - BROD NORTH MARSHALLING YARD in Yugoslavia. Eight (8) P-38's and 4 P-51's escorted the formation from start to withdrawal. The four spare aircraft returned to base. There were no early returns.

The overcast decreased in density enroute to the target with practically clear weather in the target area. Twenty-eight (28) aircraft were over the target at 1157 hours and 26 aircraft dropped 64.75 tons of 500 lb. RDX bombs from 24,000 feet by visual means. Results: Ten to fifteen hits in the yard and several hits in the locomotive factory adjacent to the marshalling yard on the north.

No enemy aircraft were seen. IAH flak was encountered at the target for 4 minutes. On return MAH flak was encountered at Sarajevo for 2 minutes. This flak appeared to come from the marshalling yard and from along the railroad, indicating the possibility of it being railroad flak. Ten (10) aircraft sustained minor flak damage. One aircraft received over 300 flak holes with damage being classified major. There were no casualties.

After receiving permission to bomb the 1st alternate target, the leader led the formation over the Adriatic and circled until an altitude of 21,000 feet was obtained before attempting to cross the coast of Yugoslavia. The overcast was very dense making navigation difficult. From Slano, the formation proceeded by dead reckoning, thence to the IP.

One aircraft is missing. Aircraft 78344 was last seen as the formation began to let down through the overcast over the Adriatic at that time the aircraft did not appear to be in any difficulty.

Weather: There were at least 10 decks of clouds from 3,000 to 20,000 feet over the base. There was 1/10 cloud coverage over the target area.

Twenty-seven (27) aircraft landed at base at 1400 hours.

829th Bomb Sqdn Aircraft 78344

| | | | |
|---|---|---|---|
| P | 2nd Lt. Arthur P. Sneden | RO | Cpl. Raymond E. Schneck |
| CP | 2nd Lt. William E. Anderson | WG | Sgt Adolphus M. Harrison |
| N | 2nd Lt. Everett G. Latham | BG | Cpl. John C. Evans |
| B | 2nd Lt. August L. Weiss | UG | Pvt Howard M. Scherr |
| ENG | Cpl. Robert C. Clasen | NG | Sgt Isaac L. Isabell |

## MISSION NO. 84 - 22 SEPTEMBER 1944

At 0824 hours, 28 B-24's and 4 spare aircraft took off to bomb OBERWEISENFELD BMW BUILDINGS (PT) in Munich, Germany. The first attack unit was led by Major O'Brien, 829th CO and the second attack unit was led by Captain Roger J. Jones, 828th Flt Cmdr. The Group assembled into Wing formation. Thirty-five (35) P-38's rendezvous with the formation at 1038 hours. At the target area, several P-51's joined the formation and remained with the formation until 1445 hours. The four spare aircraft returned to base and aircraft 504 and 750 returned early.

Twenty-six (26) aircraft were over the target at 1244 hours and dropped 64.75 tons of 500 lb. RDX bombs by PFF. The target was covered by 6/10s clouds based at 10,000 feet and partially obscured by a smoke screen. The bombs fell in an area approximately 3 miles east and southeast of the briefed IP with 10 to 15 strikes in the railroad workshop area are 3 strikes in the railroad siding. No enemy aircraft were seen. IAH flak was encountered for 5 minutes. Some red bursts of flak were observed in the target area with some white flak bursting above the formation.

Ten (10) aircraft received minor flak damage. Aircraft 438 had its oxygen system shot out on the bomb run. Two crew members passed out temporarily, from lack of oxygen before the emergency supply could be administered. There were no casualties.

Twenty-six (26) aircraft were over the field at 1545 hours. Aircraft 438 landed with no. 4 engine feathered and left tire flat, blocked the runway for 30 minutes, while the tire was changed, as the formation circled the field during this time. Five aircraft left the formation because of fuel shortage and landed at Pantenella. Twenty-one (21) aircraft landed at 1612 hours. The others returned later.

Weather: There was a 2500 foot ceiling over the base.

## MISSION NO. 85 - 23 SEPTEMBER 1944

At 0749 hours, 28 B-24's and 4 spare aircraft took off to bomb the LATISANA RAILROAD BRIDGE (PT) in northern Italy. The 1st attack, unit was led by Lt. Col. Tomhave, Gp CO and the 2nd attack unit was led by Major Dale S. Seeds, Acting 55th Bomb Wing Ops Officer. The Group assembled into Wing formation. The 3 spare aircraft flew with the 2nd unit and one returned at 0845 hours. Aircraft 885, piloted by Bryson, crash landed on the home field at 0845 hours. Eleven (11) crewmembers escaped without injury. Captain Bryson's aircraft lost no. 3 engine before the Group rendezvous and was forced to turn back. On the return, the oil pressure of no. 4 engine dropped to about 30 psi. No. 3 engine, was feathered which operates the hydraulic system, therefore, the emergency hydraulic pump had to be used. The ceiling at the time of the landing was made did not exceed 500 feet. This necessitated a low approach. The handle for lowering the landing gear was placed in the down position about two miles northwest of the field. The main gear came down and locked. However, the nose wheel did not come down and although the pilot knew that the nose gear was not down, he did not attempt to go around because of heavy gas and bomb loads. He was also expecting to lose no. 4 engine at any moment. The aircraft hit the runway, the main gear absorbing the initial shock before burning. When the gear collapsed, the aircraft spun around on the runway and skidded off into the clear area, coming to rest in a reverse position. The aircraft was demolished - bombs and pieces of the aircraft were scattered for 500 feet. No damage was done to the runway. All crew members escaped without injury. Both engines were checked and internal failure was found.

Thirty-one (31) aircraft were over the target at 1107 hours. Thirteen (13) aircraft dropped 37.5 tons of 1000 lb. GP bombs from 14,000 feet. Seventeen (17) aircraft were over the 2nd Alternate Target at 1133 hours and dropped 42.5 tons of 1000 lb. GP bombs from 12,400 feet. There were no early returns. The 1st unit bombed the primary target. The 2nd unit bombed the 2nd Alternate, the PIAVE/SDONA DI PIAVE RAILROAD BRIGE in northern Italy. Results: Undetermined on the primary target, because of a cloud coverage and the target was bomb with the aid of PFF. The 2nd unit bombed visually and hit the approach span and the east span of the bridge. No enemy aircraft were seen. Three bursts of flak were noted by aircraft of the first unit as they rallied from the primary target. No aircraft were damaged and there were no casualties.

Weather: There were 6-8/10s cloud coverage over the base of the clouds at 500 feet.
Thirty-one (31) aircraft landed at 1335 hours.

## MISSION NO. 86 - 24 SEPTEMBER 1944

At 0820 hours, 32 B-24's took off to bomb the SALONIKA WEST MARSHALLING YARD (PT) in Greece. The 1st attack unit was led by Major Reeve, Gp Ops officer and the 2nd attack unit was led by Captain Narsden D. Kelly, 831st Flt Cmdr. The Group rendezvous with the 460th BG. There was no fighter escort. Several P-38's and P-51's were observed in the target area. There were no early returns.

Thirty-two (32) aircraft were over the target at 1114 hours and 31 aircraft dropped 74.5 tons of 500 lb. GP bombs. Results: There were 45 to 50 strikes in the marshalling yard and three to four direct hits on loading sheds adjacent to the yard.

No enemy aircraft were seen. IAH flak was encountered over the target for 3 to 4 minutes. Flak was extremely accurate with the first burst hitting the lead box. Aircraft 50775 flying in no. 2 position of Able box, received a direct hit on no. 3 engine as bombs were away. The aircraft caught fire and lost altitude rapidly Before any crew members had a chance to jump, the aircraft disintegrated, breaking into three separate pieces, each of which was in flames.

Aircraft 78330, flying in no. 5 position in Able box, apparently got into the flames from aircraft 50775. The aircraft was observed to go into a steep dive in order to miss the falling aircraft ahead. One wing tip was observed in flames. The aircraft continued in the dive for several thousand feet. Two chutes were observed in the area, believed to have been from this aircraft. The aircraft pulled out of the dive and leveled off, apparently under control. The aircraft was not seen after the formation had left the target. Nine (9) aircraft received minor flak damage and three aircraft received major flak damage. Aircraft 10564 landed at Bari, Italy, to discharge two wounded crew members. The aircraft remained for repairs. Aircraft 29498 landed at Bari to discharge wounded crewmembers and returned to the base later. Bombardier, 2nd Lt. Gaylord C. Starin was wounded in both legs. The Navigator administered first aid to the wounded man during the flight, and wrapped him in flight jackets in attempt to prevent shock. The co-pilot of aircraft 460, 1st Lt. Clifford G. Studaker received a minor abrasion on the left leg from flak.

The weather was clear over the base and the target area.
Twenty-eight (28) aircraft landed at base at 1331 hours.

| 829th Bomb Sqdn Aircraft 78330 | | 829th Bomb Sqdn Aircraft 50775 | |
|---|---|---|---|
| P | 2nd Lt. James B. Cameron | P | 1st Lt. Robert E. Hegmann |
| CP | Alex E. Vroblezly, FO | CP | 1st Lt. Bryson W. Watts |
| N | 2nd Lt. William R. Meeks | N | 1st Lt. Marvin R. Weiner |

| | | | |
|---|---|---|---|
| B | 2nd Lt. William P. McLean | N | 2nd Lt. Lt. Everett G. Latham |
| ENG | SSgt Robert V. Burling | B | 1st Lt. Joe B. Hackler |
| RO | Cpl. Homer E. Jones | ENG | TSgt Walter K. Stone |
| UG | SSgt Wilson F. Leon | RWG | TSgt Thomas L. White |
| TG | Cpl. Reginal R. Lyons | NG | SSgt Joseph M. Cullen |
| BG | Cpl. Edward J. Ezakoczi | TG | SSgt Dale E. Morrison |
| NG | Cpl. Orville (nmi) Kinsberg | BG | SSgt Cecil W. Smith |

# UNIT HISTORY, 485TH BG (H) 1 OCTOBER 1944 TO 31 OCTOBER 1944

The 485TH BG (H) flew 9 operational missions during the month, making a total of 95 missions as of 31 October 1944. 630.05 tons of bombs were dropped during the month. Results were very good. Several missions during the month, where bombing was done by PFF equipment, the results were not observed.

Two (2) aircraft and their crews were lost in combat during the month. One of these crews bailed out over Yugoslavia and were evacuated safely to the base. Several other crews were reported missing and had landed safely at friendly fields, located in northern Italy and the Isle of Vis.

During the month, the following awards were given to members of this command:

| | |
|---|---|
| DFC's | 3 |
| DSC's | 1 |
| Soldiers Medal | 1 |
| Bronze Star | 1 |
| Purple Hearts | 25 |

The morale of the Group continued to maintain its high standard. The personnel have done as much as possible with the materials available to complete the winterization of their quarters. Most of the tents and the houses now have stoves of one sort or another made from anything that will serve the purpose. Each model has improved over the one that some other person designed. It is surprising what the personnel have done in improvising these, considering that the supplies and materials that can be obtained are very little and not very much to start from.

The Group theatre is now in operation. It was completed the later part of the month, after some difficulties. There are two shows nightly, so that everyone can attend, if they have the desire to do so.

## MISSION NO. 87 - 4 OCTOBER 1944

At 0855 hours, 46 B-24's, (48 scheduled) took off to bomb the ADIGE/MEZZOCORONA and the ADIGE/ORA RAILROAD BRIDGES in northern Italy. The 1st attack unit was led by Lt. Col. Herblin, Deputy CO and the 2nd attack unit was led by Major Monroe, 828th Ops Officer. Aircraft 504 and 394 did not take off. The Group rendezvous with the 460th BG. There was no fighter escort. Aircraft 638 and 371 returned early.

Twenty-one (21) aircraft of the 1st unit were over Adige Mezzocorona at 1312 hours and 20 aircraft dropped 50 tons of 1000 lb. RDX bombs. Mountains obscured the target and no strikes were made on the primary target. No flak was encountered over Adige/Mezzocorona.

Twenty-three (23) aircraft of the 2nd unit were over the Adige/Ora bridge at 1310 hours and 22 aircraft dropped 52 tons of 1000 lb. RDX bombs by visual means. Two or three direct hits were made on the bridge and several strikes on both approaches. At Adige/Ora SAH flak was encountered for 1 minute from which 4 aircraft received minor flak damage.

No enemy aircraft were seen. There were no casualties.

Weather: There was a 2-4/10 cloud coverage over the base at 10,000 feet and over the targets. Forty-four (44) aircraft returned to base at 1558 hours.

## MISSION NO. 88 - 7 OCTOBER 1944

At 0935 hours, 47 B-24's took off to bomb the VIENNA WINTERHAFFEN OIL STORAGE (PT) in Austria. The 1st attack unit was led by Lt. Col. Richard Griffin, 830th CO and the 2nd attack unit was led by Major Atkinson, 831st CO. The Group rendezvous with the 460th BG. Sixty-five (65) P-38's and P-51's escorted the formation from 1244 to 1530 hours. Eight (8) aircraft returned -prior to bombing: 78394, 40818, 51277, 78416, 32728, 78495, 78474, and 29504.

Thirty-nine aircraft were over the target at 1336 hours and dropped 96 tons of 500 lb. RDX bombs from 23,000 feet. Results: Several oil storage tanks received direct hits as well as oil barges, tank cars and several on rail sidings. No enemy aircraft were seen. IAH flak was encountered for 4 to 6 minutes. Twenty-one (21) aircraft received minor flak damage and 7 aircraft received major flak damage with over 100 holes in each. Casualties: 2nd Lt. Dean A. Davis, Navigator on aircraft 138; Sgt. George Ureszel on aircraft 508 and Lt. Todd on aircraft 299.

Lt. Col. Griffin, "A" Group leader, was forced to relinquish the lead on the rally, because of necessity for feathering one engine and a second engine was smoking badly. The lead was taken over by the deputy leader. Aircraft 912 received a direct hit in the left wing over the target and was last seen losing altitude rapidly in the target area. No chutes were definitely reported from this aircraft, although at least three chutes were observed in the target area. Aircraft 138 landed at Bari to discharge wounded crew member and returned late. Aircraft 508 landed at Foggia to discharge wounded crew member and returned late.

Weather: There was an overcast over the base and clear in the target area.

Thirty-six (36) aircraft landed at base at 1640 hours.

830th Bomb Sqdn Aircraft 912

| | | | |
|---|---|---|---|
| P | 2nd Lt. Oscar J. Rambeck | WG | Sgt. Dean R. Sasse |
| CP | 2nd Lt. Raymond G. Johnson | NG | MSgt. Milton L. Emery |

| | | | |
|---|---|---|---|
| N | 2nd Lt. Theodore J. Merchant | UG | Sgt. John T. Julenics |
| B | 2nd Lt. Glenn G. Strom | TG | Sgt. Richard J. Dawe |
| WG | Sgt. John C. Von Husen | BG | Cpl. Jesse C. Harvey |

Aircraft received a direct flak hit which knocked the left wing tip off completely just before bombs were away. No chutes were seen from this aircraft.

## MISSION NO. 89 - 10 OCTOBER 1944

At 0841 hours, 28 B-24's and 4 spare aircraft took off to bomb the PIAVE/ SUSEGANA RAILROAD BRIDGE in northern Italy. The 1st attack unit was led by Lt. Col. Nett, 828th CO and the 2nd attack unit was led by 1st Lt. Lloyd Allen, 829th Flt. CMdr. There was no rendezvous with other Groups. There was no fighter escort, however, several P-38's were observed over the target area. There were no early returns.

Thirty-two (32) aircraft were over the target at 1128 hours at 18,000 feet altitude. A complete undercast obscured the target and the IP, making it impossible to do precision bombing. It was decided to return to base rather then to bomb indiscriminately. No enemy aircraft were seen. SAH flak was encountered over the target for 2 minutes with three aircraft receiving minor flak damage.

Weather: The were 7/10s altostratus clouds over the base and 8/10s altocumulus clouds over target area.

Thirty-two (32) aircraft landed at base at 1407 hours. Thirty-one (31) aircraft returned 310 500 lb. RDX bombs and one aircraft jettisoned 10 bombs to lighten its load after having feathered no. 4 engine.

## MISSION NO. 90 - 12 OCTOBER 1944

At 0742 hours, 36 B-24's, scheduled, and took off to bomb an AMMUNITION DUMP at Bologna, Italy. The 1st attack unit was led by Major O'Brien, 829th CO and the 2nd attack unit was led by Captain Roger A. Nichols, 828th Flt. Cmdr. The Group did not assemble into Wing formation. Nine (9) P-38's escorted the formation from 1153 to 1253 hours. There were no early returns.

Thirty-six (36) aircraft were over the target at 1127 hours and 24 aircraft dropped 46.2 tons of 1000lb. GP bombs. Because of a malfunction of bombing mechanism, 12 aircraft returned their bomb load to the base. No military damage accomplished. SAH flak was encountered for minutes over the target. Eighteen (18) aircraft received minor flak damage. Aircraft 422 received major flak damage to no. 3 engine. Aircraft 730 had its hydraulic system shot out. There were two casualties: TSgt F. W. Chapman and Sgt. Jack Peters. No enemy aircraft were seen.

Weather: Over the base there were 1/10 altocumulus clouds and it was clear over the target area.

Aircraft 730, having its hydraulic system damaged by flak, had the right landing gear to collapse on landing. The pilot made an excellent landing, keeping the aircraft under control which skidded to a stop off the runway without injury to any of the crew members.

Thirty-five (35) aircraft landed at base at 1354 hours.

## MISSION NO. 91 - 12 OCTOBER 1944

At 0809 hours, 37 B-24's, scheduled, and 2 spare aircraft took off to bomb the BANHIDA SOUTH MARSHALLING YARD (PT) in Hungary. The 1st attack unit was led by Lt. Col. Herblin, the 2nd attack unit was led by Major Monroe, 828th Ops Officer and the 3rd attack unit was led by Major John B. Stoddart, 830th Ops Officer. The Group assembled with the 460th BG. Ten (10) P-38's escorted the formation from 1007 hours to 1315 hours. The two spare aircraft left the formation and returned to base. There were no early returns.

Thirty-six (36) aircraft were over the target at 1136 hours at 17,000 feet and 24 aircraft dropped 59 tons of 500 lb. RDX bombs. Twelve (12) aircraft dropped 30 tons of 500 lb. incendiaries. Bombing was accomplished by visual means. The bomb pattern scattered with bomb concentrations west of the marshalling yard. Twenty (20) strikes were in the south end of the marshalling yard. No enemy aircraft were seen. SAH flak was encountered at the target for 1 minute. There was some small caliber flak, including tracers, observed bursting below the formation. Seven (7) aircraft received minor flak damage. There were no casualties.

Weather: it was clear over the base and the target area.

Thirty-six (36) aircraft landed at the base at 1444 hours.

## MISSION NO. 92 - 16 OCTOBER 1944

At 0716 hours, 38 B-24's (40 scheduled) took off to bomb a primary target in Austria. The 2nd resort target, GRAZ NEUDORF AIRCRAFT ENGINE FACTORY in Austria, was bombed after an unsuccessful attempt had been made to bomb the PT. The 1st attack unit was led by Lt. Col. Tomhave; the 2nd attack unit was led by 1st Lt. Lloyd Allan, 829th Flt Cmdr and the 3rd attack unit was led by 2nd Lt./ Donald L. Gambrill, 830th pilot. The Group assembled into Wing formation. Thirty-five (35) P-38's escorted the formation from 1015 to 1331 hours. Aircraft 335 and 725 returned early.

Thirty-six (36) aircraft were over the primary target and started a bomb run. Both the IP and the PT were completely obscured by an undercast. Faulty PFF equipment made bombing of the PT impossible. Two aircraft left the formation due to engine trouble; Aircraft 486 and 750. Thirty-four (34) aircraft were over the alternate target at 1207 hours and 31 aircraft dropped 77.5 tons of 500 lb. RDX bombs. The target was obscured by clouds and smoke from bombing by the 460th BG. Crews reported scattered pattern with some hits on building southwest of the center building which was the briefed MPI. No enemy aircraft were seen. S-MAH flak was encountered over Graz for 2-3 minutes. Twelve (12) aircraft received minor flak damage.

Aircraft 394 and 872 landed at Ancona for fuel and returned later. Aircraft 067 landed at Ancona for fuel and remained over night. Aircraft 384 landed at Biferno for fuel and returned later. Aircraft 559 is believed to have ditched in the Adriatic. Aircraft 271 is missing.

Thirty (30) aircraft landed at base at 1511 hours.

829th Bomb Sqdn Aircraft 559

| | | | |
|---|---|---|---|
| P | 1st Lt. Richard H. Boehme | RO | Cpl. Abe (NMI) Goldman |
| CP | 2nd Lt. Mervin W. Jacobson | BG | Cpl. Harold L. Oliver |
| N | 2nd Lt. Bob B. Bishop | UG | SSgt Billy R. Culver |
| B | 2nd Lt. Richard C. Weintritt | TG | Cpl. Kenneth H. Lohr |
| ENG | Cpl. Robert E. Brown | | |

Aircraft was heard to call BIG FENCE -- later MAYDAY. Thought to have ditched, as bombardier was heard over command radio to order crew to jettison guns.

## MISSION NO. 93 - 17 OCTOBER 1944

Marshalling Yard - Nagykaniza, Hungary
NO RECORD RE ABOVE MISSION IN HISTORY

## MISSION NO. 94 - 20 OCTOBER 1944

At 0833 hours, 42 E-24's, scheduled, and took off to bomb a primary target in Germany. Due to adverse weather conditions the PT could not be located and the ROSENHEIM MARSHALLING YARD in Germany (1st Alternate target) was bombed. The 1st attack unit was led by Lt. Col. Griffin, 828th CO; the 2nd attack unit was led by Major Monroe, 828th Ops Officer and the 3rd attack unit was led by Captain Roger Jones, 831st Flt Cmdr. "D" box after taking off was unable to locate the formation and proceeded on course alone. Wing formation was not accomplished due to weather conditions. Eleven (11) P-38's escorted the formation from 1155 to 1400 hours. Eleven (11) aircraft returned prior to bombing: Aircraft 750, 827, and 752 returned because of mechanical troubles. "D" box - 508, 819, 699, 872, 394, 834 and 829 - being unable to find the formation, returned to base at 1415 hours. All bombs were returned.

Thirty-one (31) aircraft were over the target at 1256 hours and 30 aircraft dropped 44.25 tons of 500 lb. RDX bombs from 21,500 feet. The target was cloud covered. The main bomb pattern lies southeast of the marshalling yard with several strikes on the railroad tracks to Kufstein, probably cutting the line. No enemy aircraft were seen and no flak was encountered. There was no battle damage or casualties.

A B-24 from the 460th Group was observed to spin, over the Adriatic. Two B-24's collided over the Adriatic from which 16 chutes were seen. Two aircraft from other Groups circled and dropped life rafts near the survivors.

Weather: There were 8/10s cumulus, stratocumulus and towering cumulus with showers over the base with 6/10s cumulus over the base on return.

Thirty (30) aircraft landed at base at 1531 hours. The last non-sortie aircraft to return had its main landing gear to collapse while taxing and blocked one taxi strip for several minutes. Aircraft 335 landed at Lesi for refueling and remained over night.

## MISSION NO. 95 - 23 OCTOBER 1944

At 0735 hours, 45 B-24's (47 scheduled) took off to bomb the AUGSBURG M.A.M. WORKS in Germany. The 1st attack unit was led by Lt. Col. Nett, 828th CO and the 2nd attack unit was led by Major Cummings, 829th CO. The Group assembled into Wing formation. Forty (40) P-38's escorted the formation from 1100 hours to 1318 hours. Aircraft 818, 394 and 299 returned prior to bombing.

Forty-two (42) aircraft were over the target at 1214 hours and 41 aircraft dropped 102 tons of 1000 lb. RDX bombs from 21,000 feet by PFF. Because of a complete undercast, it was estimated that the bombs fell slightly left of the target and in the center of the city. No enemy aircraft were seen. MAH flak was encountered for 4 minutes in the target area. Weather conditions necessitated a deviation in the course after the rally which carried the formation over the westerly edge of Innsbruck where M-IAH flak was encountered. Evasive action limited the exposure to approximately 1 minute. Eleven (11) aircraft received minor flak damage. There were no casualties.

Weather: Over the base and assembly area there were 3-5/10s stratocumulus and cumulus with showers - bases of clouds at 3,000 feet.

Thirty-six (36) aircraft landed at 1552 hours at base. Aircraft 438 landed at Foggia Main for refueling and returned later. Aircraft 422 landed at Bari for refueling and returned later.

Aircraft 801 landed at Lesi airdrome; 416 landed at Falconara airdrome and 725 landed at Fano airdrome. Aircraft 446 is missing. Last reported with no. 3 engine feathered at 1242 hours and dropped out of formation with the aircraft under control.

# UNIT HISTORY, 485TH BG (H) 1 NOVEMBER 1944 TO 30 NOVEMBER 1944

The 485th BG (H) flew 16 operational mission during the month, making a grand total of 111 missions as of 30 November 1944. 595.2 tons of bombs were dropped during the month. There was a decrease in tonnage due to fewer normal effort missions. The results on observed missions were generally good. This, also, was the first month that PFF equipped aircraft were used to bomb individually. Also, the Groups first night mission was flown on the 24-25 of November. Propaganda bombs were first used by the Group on the 17th of November 1944, with the Blechhammer South Oil Refinery in Germany as the target. Delayed action bombs were also used on this mission.

Three (3) aircraft and their crews were reported missing but later were verified to have landed at friendly airfields.

No enemy aircraft were encountered during the month.

| | |
|---|---|
| Aircraft assigned - 1 Nov 1944 | 64 |
| Aircraft assigned - 30 Nov 1944 | 48 |
| Combat losses – 1-30 Nov 1944 | 4 |
| Non-combat losses – 1-30 Nov 1944 | 3 |

Fourteen (14) new crews were received during the month.

The winterization program continued throughout the month with increased building activity. Many tufa blockhouses were completed and practically all tents were given some form of winterization. Heaters, improvised from salvaged materials were installed in the quarters of all personnel, and dumps for fuel oil were established. MUD has been prevalent during the entire month. In an effort to get out of the mud, walks were constructed; company streets and parking areas for vehicles were designated in all Sqdn areas. On 23 Nov, the Group theatre was formally opened, with a seating capacity of 1150.

AWARDS

| | |
|---|---|
| Legion of Merit | 2 |
| DFC | 12 |
| Bronze Star | 1 |
| Purple Hearts | 6 |

## MISSION NO. 96 - 1 NOV 1944

At 0900 hours 28 (and 3 spares) scheduled B-24s took-off to bomb the primary target. The 2nd alternate target - marshalling yard at Graz, Austria - was bombed because the primary target was covered by cirrus clouds with bases of 18,000 feet and tops well over 24,000 feet. The formation reached a point 15 miles south of the primary target's IP before the decision was made to bomb the 2nd alternate target. The 1st attack unit was led by Major Atkinson, 831 CO and the 2nd attack unit was led by 1st Lt. Eugene B. Lenfest, 830 Flt Cmdr. The three spare aircraft returned to base.

The Group assembled into Wing formation. Thirty to thirty-five P-51's escorted from 1238 hours to 1538 hours. Aircraft 801 and 426 returned prior to bombing.

Twenty-two aircraft were over the target at 1338 hours and dropped 55 tons of 500 lb. RDX's. "B" box slid over and above "C" box, which prevented the leader of the 2nd attack unit from flying a straight course. This was brought about by the dense cirrus cloud formation, which made visual observation of adjacent boxes difficult. Three aircraft became separated from the formation in the cirrus clouds near the primary target's IP and bombed a target of opportunity - marshalling yard at Komend - dropping, 7.5 tons Of 500 lb. RDX's. Also, 067 became separated from the formation and bombed the town of Studenzen, dropping 2.5 tons of 500 lb. RDX's from 19,000 feet at 1301 hours. Results of bombing at Graz: Bomb pattern located short of the marshalling yard. No military damage was noted.

The weather at the base was 3/10 coverage with bases at 6,000 feet.

No enemy aircraft were seen. Scant to moderate - intense to Accurate - Heavy flak was encountered in the target area for 4 minutes. Seven aircraft received minor flak damage. There were no casualties.

067 ran low on gas after having become separated from the formation and landed at the airdrome north of Ancona, Italy - refueled and returned to base. Twenty-five aircraft landed at 1612 hours.

## MISSION NO. 97 - 3 NOV 1944

Three (3) B-24's: Aircraft 52034 piloted by Lt. Col. Herblin, Deputy Gp CO; aircraft 48758 piloted by 1st Lt. Joseph W. Cathey, 830th and aircraft 48779 piloted 2nd Lt. William R. Fritz, 828th took off to bomb the MUNICH WEST MARSHALLING YARD (PT) in Germany. Scheduled take off was delayed 17 minutes by local thunderstorms, but the aircraft arrived in the Mafredonia - Bari rendezvous area at 7500 feet altitude well within the scheduled time. They departed on course from this rendezvous area, leaving at 1-minute intervals, starting at 0855 hours. There was no assembly with other Groups and there was no fighter escort.

No enemy aircraft were seen. Aircraft 52034 was in IIH flak for 4 minutes with the flak bursting low and behind the aircraft. Aircraft 48758 in IAH flak for 4 minutes with the flak bursting slightly behind the aircraft. Aircraft 48779 was in IAH flak for 3-4minutes.AllAircraft bombed by PFF. 4.5 tons of 250 lb. GP bombs were dropped by the three aircraft. Operators indicated that the bombs fell in the target area or just over in the city proper.

All three aircraft returned to base between 1433 and 1450 hours lately.

## MISSION NO. 98 - 4 NOV 1944

At 0740 hours, 36 B-24's, scheduled, and 2 spare aircraft took off to bomb the LINZ BENZOL PLANT (PT) in Austria. The 1st attack unit was led by Major O'Brien, 829th CO, the 2nd attack unit was led by 1st Lt. Charles E. Porter, Jr., 830th Flt Cmdr and the 3rd attack unit was led by Major Calvin W. Fite, 828th Ops Officer. The Group assembled into Wing formation. Twenty-five (25) to thirty-five (35) P-38's escorted the formation from 1055 to 1320 hours. Ten P-51's were observed on withdrawal from the target. The 2 spare aircraft returned to base plus ten aircraft returned prior to bombing: 899, 043, 728, 425, 438, 931, 067, 819, 927 and 727.

Twenty-six (26) aircraft were over the target at 1151 hours and dropped 63.75 tons (52 tons of 500 lb. RDX bombs and 11.75 tons of 500 lb. RDX bombs) by PFF. The target was completely covered with clouds. The formation was crowded to the left on the bomb run because the rear unit of the Group ahead was flying at the same altitude. The PFF noted that the target could not be hit so bombs were released to strike in the city to the north and west of the briefed target. No enemy aircraft were seen. MIH flak was encountered over the target for 5 minutes. Flak was low and to the right with probability that fire was aimed at other Groups in the area. No aircraft were damaged and there were no casualties.

Weather: There were a few stratocumulus over the base. Over the target there were 9/10s altocumulus at 17,000 feet. Temperature at 24,000 feet was -35 degrees.

Aircraft 157 landed at Foggia for refueling and returned later. Aircraft 498 landed at Lesi with aircraft 299. Aircraft 791 is missing. No reports are available on its separation from the formation, other that it was last seen at the target.

Twenty-two (22) aircraft landed at 1505 hours.

## MISSION NO. 99 - 4 NOV 1944

At 1220 hours, 2 B-24's took off to bomb PODGORICA (PT) in Yugoslavia. Lead aircraft was piloted by Lt. Col. William L. Herblin, Deputy Gp CO and the other aircraft was piloted by Major Reeve, Gp Ops Officer. There was no rendezvous with other Groups. Rendezvous with 4 P-38's was effected at base at 1220 hours. The escort left the two bombers at the Italian coast on the return trip.

Two (2) aircraft were over the target at 1349 hours and dropped 4.5 tons of 500 lb. RDX bombs by visual means. Results: The selected building was hit with concentration of bombs on the southwest section. Target was largest building in the city. No enemy aircraft were seen and no flak was encountered. There were no casualties.

Weather: The route was CAVU with 5-6/10s cumulus clouds over the target.

The two aircraft landed at base at 1453 hours.

## MISSION NO. 100 - 5 NOV 1944

At 0807 hours, 27 B-24's took off to bomb the FLORISDORF OIL REFINERY at Vienna, Austria. The 1st attack unit was led by Lt. Col. Tomhave, Gp CO and the 2nd attack unit was led by Captain Roger J. Jones, 828th Flt Cmdr. The Sqdns assembled over the base and proceeded to Altamura where the Group assembled at 0848 hours. Aircraft 1299 collided in mid-air with Aircraft 394 at 0845 hours approximately 30 miles south of Venosa Airdrome. Aircraft 1299 crashed, killing 8 out of 10 crewmembers. Aircraft 394 returned to base at 0900 hours. Wing formation was completed at 0902 hours. Forty-five (45) to fifty (50) P-51's escorted the formation from 1023 to 1308 hours. Several P-38's were observed at the target area. Aircraft 394, 694 and 699 returned prior to bombing.

Twenty-three (23) aircraft were over the target at 1219 hours. Nine (9) aircraft dropped 18 tons of 500 lb. RDX bombs and 12 aircraft dropped 30 tons of incendiaries. Bombing was accomplished by PFF. Results were undetermined due to complete cloud coverage of the target area. Two Me 109's were seen near the IP but did not attack the formation. IIH - IAH flak was encountered over the target for 6 minutes. Four (4) aircraft received minor flak damage. There was one casualty: SST Kerri, 828th gunner.

Weather: There were 7-8/10s cumulus and stratocumulus at 3000 feet over the base.

Twenty-two (22) aircraft landed at base at 1455 hours. Aircraft 043 landed at Bari with wounded crewmember and returned later.

## MISSION NO. 101 - 5 NOV 1944

At 1012 hours, 6 B-24's, scheduled, took off to bomb PODGORICA in Yugoslavia. The flight was led by Captain Marsden G. Kelly, 831st Flt Cmdr. No rendezvous was effected with other Groups or fighter escort.

Six (6) aircraft were over the target at 1148 hours and dropped 15 tons of 500 lb. GP bombs from 18,000 feet. The 460th BG was seen in the target area. Bombing was accomplished by visual means. Results: The bomb pattern was in the south central portion of the city. Several blocks of buildings were damaged from direct hits and near misses. Several direct hits were made on the largest building in the city. No enemy aircraft were seen and no flak was encountered.

Weather: Enroute and on the return there were scattered stratocumulus. Over the target area it was CAVU.

Six (6) aircraft landed at 1251 hours.

## MISSION NO. 102 - 6 NOV 1944

At 0714 hours, 31 scheduled B-24's took-off to bomb the primary target - VIENNA SOUTH ORDNANCE DEPOT. The 1st attack unit was led by Lt. Col. Herblin, Deputy Gp CO and the 2nd attack unit was led by Captain Eugene B. Lenfest, 830 Flt Cmdr.

The Group assembled into Wing formation. Thirty-five P-38's escorted the Group from 0827 hours to 1300 hours. Aircraft 495 and 908 returned to base prior to bombing.

The weather ship was contacted enroute and it reported that the primary and all alternate targets were cloud covered. 10/10 cloud coverage based at 8,000 feet prevented bombing by visual means. Being unable to identify the IP by PFF, the formation turned on the ETA of the IP and made the bomb run on PFF.

Twenty-nine aircraft were over the target at 1121 hours at 21,000 feet and 17 aircraft dropped 23.5 tons of 500 lb. RDX's; 9 aircraft dropped 22.5 tons of 500 lb. cluster incendiaries.

Four aircraft dropped 8.5 tons of 500 lb. RDX's short of the target due to accidental release. The bombs were scattered over a large area as several aircraft dropped late when the leader of the 2nd attack unit held it's bombs for several seconds believing that the 1st attack unit had erred in target identification. After bombs were released, the leader of the 460th Bomb Group was contacted and verification of the target was given by the Group's PFF operator. Because of cloud coverage it was impossible to determine the exact location of the bombing.

No enemy fighters or flak was observed.

The weather was 2/10 to 8/10 over the base and the rendezvous area. The target area was completely covered with stratocumulus at 8 to 10,000 feet.

Return was made without incident and 28 aircraft landed at 1352 hours. One aircraft 718 - pilot Gigowski - landed at Vis and is there at the time of the writing of this report.

## MISSION NO. 103 - 7 NOV 1944

At 0905 hours, 28 B-24's, scheduled, and one spare aircraft took off to bomb the ADIGE/ORA RAILROAD BRIDGE in northern Italy (PT). The 1st attack unit was led by Major Reeve, Gp Ops Officer and the 2nd attack unit was led by 1st Lt. Charles E. Porter, 830th Flt Cmdr. The Group assembled into Wing formation. Aircraft 498 returned early. There was no rendezvous with fighter escort.

Twenty-seven (27) aircraft were over the target at 1306 hours and dropped 66.25 tons of 500 lb. RDX bombs by visual means. Results: One bomb pattern was from two boxes, which fell at the north end of the bridge, probably damaging the approach and the first span. Two patterns fell approximately 5,000 feet west of the bridge.

Due to an overcast with bases at 17,000 feet the formation bombed from 16,000 feet. The flak defenses at the target seemed to have the range of the formation perfectly and effectively placed their burst in each box. Another factor, which heavily favored the flak defenses, was a strong head wind which appreciably slowed the formation. IAH flak, supplemented by accurate medium flak, was experienced for 4 minutes. Immediately after bombs away, the lead aircraft of the formation, being damaged by flak, relinquished the lead position to the deputy lead aircraft, piloted by Captain Roger J. Jones, Flt Cmdr. Charley box reported being in flak intermittently for approximately 30 minutes from the target to the coast on return. No enemy aircraft were see.

Eight (8) aircraft received major flak damage and landed at friendly fields. Aircraft 064, pilot Major Reeve, landed at Bari with wounded navigator; Aircraft 536, pilot Rasco, landed at Gioia; Aircraft 915, pilot Nikis, landed at Bari with wounded engineer and radio operator; Aircraft 507, pilot Fraser, landed at Foggia. with two wounded men; Aircraft 508, pilot Brady, landed at Iesi; Aircraft 728, pilot Schueler, landed at Gioia; Aircraft 764, pilot Schackleford, landed at Bari with wounded gunner; Aircraft 801, pilot Mason, landed at Falconara. Aircraft 764 and 915 returned later to base. In addition to the above, 15 aircraft received minor and semi-major damage from flak. Only 4 aircraft escaped flak damage.

Weather: There were 1/10 stratocumulus with bases at 4,000 feet and 6/10s cirrostratus over the base and it was clear over the target area below 18,000 feet.

Nineteen (19) aircraft landed at 1556 hours.

## MISSION NO. 104 - 15 NOV 1944

From 0804 to 0816 hours, 4 B-24's took off to bomb the LINIZ BENZOL OIL REFINERY (PT) in Austria: Aircraft 48779, pilot Lt. Col. Herblin, Deputy CO; Aircraft 48758, pilot 1st Lt. Arthur W. Rasco, 830th; Aircraft 52034, pilot 1st Lt. Christian H. Schaefer, 831st; Aircraft 51969, pilot 1st Lt. Charles H. Fabry, 828th. The aircraft proceeded to Giovinozzo, leaving that point on course at 8,000 feet at two-minute intervals, beginning at 0853 hours. There was no rendezvous with other Groups and no fighter escort. No enemy aircraft were seen.

Aircraft 48799 dropped 2 tons of 500 lb. RDX bombs from 26,000 feet by PFF and was exposed to S-MIH flak for 3 minutes, with the flak bursting below and behind the aircraft.

Aircraft 52034 dropped 2 tons of 500 lb. RDX bombs from 24,640 feet by PFF and was exposed to S-MIH flak for 4 minutes, with flak bursting low and behind and on either side of the aircraft. Aircraft encountered MIH flak for 2 minutes at Salzburg, Austria.

Aircraft 48758 turned back at 1042 hours because of weather conditions.

Aircraft 51969 is missing and disposition of its bombs is unknown.

All bombs fell in the general target area. No casualties from flak.

Weather: There was 3-4/10s cloud coverage over the base on take off. The target area was cloud covered. Clouds became a solid overcast over the Alps with severe turbulence. Over the target there was a solid undercast to 25,000 feet. Clouds decreased further south with 6-8/10s over the base.

Aircraft 779 landed at 1455 hours. Due to a very strong cross wind aircraft 52034 landed at Pantanella at 1450 hours as did aircraft 48758.

828th Bomb Sqdn Aircraft 51969

| | | | |
|---|---|---|---|
| P | 1st Lt. Charles H. Fabry | G | SSgt Robert N. Halihan |
| CP | 2nd Lt. William H. Turner | G | Sgt. Charles J. Gonsior |
| N | 2nd Lt. Harold E. Daniels | G | Sgt. Frederick F. La Plante |
| PFFN | 1st Lt. Raffaele F. Biondi | G | Sgt. Joseph J. Wallace |
| B | 2nd Lt. Emil (nmi) Opalka | | |
| G | Sgt. Robert B. Prentiss | | No information on crew. |

## MISSION NO. 105 - 16 NOV 1944

At 0803 hours, 28 B-24's and one spare aircraft took off to bomb the MUNICH WEST MARSHALLING YARD (PT) in Germany. The 1st attack unit was led by Lt. Col. Griffin, 830th CO and the 2nd attack unit was led by 1st Lt. Gerald. H. Morris, Jr., 831st Flt Cmdr. The spare aircraft returned to base. The Group assembled into Wing formation. Thirty (30) to forty (40) P-51's passed the formation over the Adriatic Sea at 1051 hours. The fighters were not seen for the remainder of the mission. Seven aircraft returned prior to bombing: 299, 633, 699, 144, 819, 486, and 899.

Twenty-one (21) aircraft were over the target at 1307 hours and 20 aircraft dropped 30 tons of 500 lb. RDX bombs from 25,000 feet, by PFF. The results were undetermined as the target was completely cloud covered.

Aircraft 48764 landed at Tortorella to refuel and returned later. Aircraft 78656 landed at Madna to refuel and returned later. Aircraft 51931 landed at Biferno to refuel and returned later. Aircraft 95460 landed at Biferno and returned later. Aircraft 52067 landed at Ancona, aircraft 78495 landed at San Salvo, aircraft 78426 landed at Falconara and aircraft 78438 landed at Falconara.

Weather: There was 6/10s thin stratus with bases at 6,000 feet over the airdrome. Vapor trails formed between 23 - 25,000 feet.

Thirteen (13) aircraft landed at base at 1615 hours. Several crews reported activity on airdromes in northern Italy. At Udine, the airfield was being repaired. Fifty revetments appeared to be in good condition. Five aircraft in revetments was noted.

## MISSION NO. 106 - 17 NOV 1944

At 0721 hours, 21 B-24's took off to bomb the SOUTH BLECHHAMMER SYNTHETIC OIL REFINERY (PT) in Germany. The two spare aircraft failed to take off. The Group formation was led by Lt. Col. Nett, 828th CO, who was forced to relinquish the lead to Major Fite, 828th Ops Officer. The Group assembled into Wing formation and left the formation with the 460th BG at 0829 hours. Thirty (30) P-38's escorted the formation from 1042 to 1401 hours. Twenty-five (25) P-51's were observed upon withdrawal from the target area. Two aircraft returned prior to bombing: 875 and 034.

Nineteen (19) aircraft were over the target at 1237 hours and dropped 28.5 tons of 500 lb. RDX bombs from 23,500 feet. The results were undetermined as the target was completely cloud covered. No enemy aircraft were seen. MIH flak, for 2 minutes, was encountered over the target area. There was no damage or casualties.

Aircraft 50899 reported, by radio, that two engines were out and was attempting to make Falconara. Arrival at Falconera has not been verified. Aircraft 78474 landed at Vis, as did aircraft 51638. Aircraft 41144 is missing. Other aircraft reported overhearing an order to bail out at 1421 hours.

Weather: There were scattered low clouds over the base.

Fifteen (15) aircraft landed at 1538 hours.

829th Bomb Sqdn Aircraft 50899

| | |
|---|---|
| P | 2nd Lt. Charles P. Stewart |
| CP | 1st Lt. Richard H. Boehme |
| B | 2nd Lt. Harry R. Carter |
| ENG | SSgt John H. Elmore |
| RO | SSgt Claud E. Martin |
| BG | SSgt Powell Robinson, Jr. |
| TG | Sgt. Richard L. Hein |
| UG | SFt Joe C. Sedlak |
| NG | Sgt. George C. Schiazza |

The crew of aircraft 908 reported that Stewart's crew to have bailed out over Yugoslavia. Co-pilot Boehme bailed out over Yugoslavia once before. Radioed that aircraft was low on gas.

## MISSION NO. 107 - 18 NOV 1944

At 0725 hours, 24 B-24's (25 scheduled) took off to bomb the UDINE AIRDROME in northern Italy (PT). The 1st attack unit was led by Lt. Col. John Atkinson, 831st CO and the 2nd attack unit was led by 1st Lt. Donald Gambrill, 830th Flt Cmdr. The Group assembled into Wing formation. There was no fighter escort. Several P-51's and Spitfires were observed over the target area. Aircraft 694 returned early.

Twenty-three (23) aircraft were over the target at 1115 hours and dropped 51.7 tons of 125 lb. clustered fragmentation bombs from 22,000 feet. Results: Two large revetments damaged. Considerable damage was done to the taxi strips and roadways south of the runway. No visual damage to aircraft in the revetments. No enemy aircraft were seen. NIH to MAH flak was encountered over the target for 3 minutes. Three aircraft received minor flak damage.

Weather: It was clear over the base at take off tire. It was clear over the target area and bombing was accomplished visually.

Twenty-two (22) aircraft landed at base at 1309 hours. Aircraft 758 landed at Foggia with wounded crew member, Sgt. Pace, and returned to base later.

## MISSION NO. 108 - 19 NOV 1944

At 0730, 24 scheduled B-24's took-off to bomb the primary target - HORSCHING AIRDROME, AUSTRIA. The 1st attack unit was led by Major O'Brien, 829th CO and the 2nd attack unit was led by Lt. Col. William L. Herblin, Deputy Gp CO.

The Group assembled into Wing formation. Twenty to twenty-five P-51's escorted the Group from 1155 hours in the target area and departed at 1230 hours. Aircraft 564 returned early.

Twenty-three aircraft were over the target at 1154 hours and dropped 50.3 tons of 125 lb. fragmentation bombs from 24,500 feet. The bomb run was visual. Results: Northeast corner of the airdrome area completely covered. Considerable damage to the east end of the north runway, east perimeter track, north perimeter track and fueling points. Enemy aircraft present in target area: 18 fighters, 4 transports; Aircraft destroyed: 6 fighters, 1 transport; Aircraft damaged: 3 fighters.

The weather over the base was clear. Target area was clear although visibility at the surface was restricted by smoke.

No enemy aircraft were encountered. Scant-inaccurate to Accurate-heavy flak was experienced for 2 to 3 minutes. Five aircraft received minor flak damage. There were no casualties. Aircraft 51932 landed at Vis to refuel and returned later to base.

Twenty-two aircraft landed at 1444 hours.

## MISSION NO. 109 - 20 NOV 1944

At 0740 hours, 27 B-24's (28 scheduled) took off to bomb the BLECHHAMMER SOUTH OIL REFINERY (PT) in Germany. The 1st attack unit was led by Lt. Col. Tomhave, Gp CO and the 2nd attack unit was led by 1st Lt. Robert L. Brown, 831st Asst. Ops Officer. The group assembled into Wing formation. Forty (40) P-38's escorted the formation, departing at 1420 hours. Several P-51's were observed over the target area. There were no early returns.

Twenty-seven (27) aircraft were over the target at 1227 hours and 26 aircraft dropped 38.5 tons of 500 lb. RDX bombs and 13 tons of 500 lb. RDX bombs by visual means. The results were excellent with direct hits on the hydrogenation stills and other installations. No enemy aircraft were seen. IAH flak was encountered over the target for 6 minutes. Four (4) aircraft received major flak damage and 12 aircraft received minor flak damage. There were no casualties.

Aircraft 932 landed at Pantanella, returning later to base. Aircraft 277 landed at Vis, returning later to base. Aircraft 829 landed at Biferno, returning later to base. Aircraft 779 landed at Torretto, returning to base later. Aircraft 446 landed at Falconara and remained for repairs. Aircraft 997 landed at Amendala. Aircraft 51872, pilot Warner, is missing. Aircraft was last seen as the formation rallied from the target, at which time it was lagging.

The first aircraft to land, 486, landed at 1600 hours with a flat tire and blocked the runway for several minutes. At 1621 hours, landings were resumed. Twenty (20) aircraft were down at 1659 hours. The last aircraft made an excellent landing with a flat nose wheel.

Weather: There were 3-6/10s cirrus at 24,000 feet.

831st Bomb Sqdn Aircraft 51872 - Blue H

| P   | 2nd Lt. Jerome P. Warner     | NG Cpl. Wilmer L. Hogan        |
|-----|------------------------------|--------------------------------|
| CP  | 2nd Lt. Joseph Drutz         | BG Cpl. Thomas R. Cook, Jr.    |
| B   | 2nd Lt. Marinus Mieras       |                                |
| ENG | Sgt. Duane G. Waters         | Blue H dropped out of formation over the target with a feathered engine. It was last seen lagging behind the Group after the rally off the target. |
| RO  | Cpl. Maurice A. Bendes       |                                |
| TG  | Cpl. Nathan Feiden           |                                |
| UG  | Cpl. Francis X. Hennessey    |                                |

## MISSION NO. 110 - 22 NOV 1944

At 0728 hours, 25 B-24's (26 scheduled) took off to bomb a target in Germany. Prevented from bombing the primary target by unfavorable weather, the 1st alternate target - SALZBURG MARSHALLING YARD in Austria was bombed. The 1st attack unit was led by Major Reeve, Ops Officer and the 2nd attack unit was led by Captain Joseph S. Gill, 828th Asst. Ops Officer. Aircraft 78299 crashed approximately 10 miles southeast of Nature airdrome at 0800 hours. Three crew members parachuted safely, the remaining seven members were killed in the crash. The Group assembled late into Wing formation due to unfavorable weather, forming over the Adriatic Sea into Wing formation. Fifteen (15) to twenty (20) P38's escorted the formation from 1154 to 1255 hours. Several P-51's were observed in the target area. Aircraft 29504, 51633 and 78474 returned early.

At the Key Point, the formation was at an altitude of 21,500 feet. Weather was clear except for patch stratocumulus. Fifty miles south of Salzburg, an altostratus overcast, based at 18,500 feet and becoming thicker to the north and west, was encountered. The formation descended to 18,000 feet to get under this front. Continuing on course the weather continued to worsen and the decision was made to bomb the 1st alternate target.

Twenty-one (21) aircraft were over the target at 1224 hours and 17 aircraft dropped 34 tons of 500 lb. GP bombs from 18,000 feet. Three direct hits were made on the industrial installations adjacent to the west marshalling yard. Many strikes were made on the roads and built up area between the west and east marshalling yard. One direct hit was made on the turntable at the south end of the west marshalling yard. Several direct hits were made on the south choke point of the east marshalling yard causing damage to the overpass.

No enemy aircraft were encountered. Two unidentified single-engine fighters were seen. Flak over the target was IAH for five minutes. Seven aircraft received minor flak damage and three aircraft received major flak damage.

The lead aircraft 52043, pilot Reeve, relinquished the lead to the deputy leader, 1st Lt. Donald L. Gambrill, 830th Flt Cmdr, aircraft 49024, at the Key Point on return. Aircraft 52043 left the formation and landed at Falconara at 1401 hours with four wounded crew members. Aircraft will return to base on 23 November 1944. Additional casualties: On aircraft 51932, Sgt. Martin; aircraft 48764, Sgt. Parsons and on aircraft 49024, Sgt. Rowe.

Weather: There were 6110s cumulus and stratocumulus clouds, with bases at 4,000 feet, over the base.
Twenty (20) aircraft landed at base at 1505 hours.

## MISSION NO. 111 - 25 NOV 1944

Two aircraft, 95517, piloted by 1st Lt. Joseph W. Cahtey, 830th and 48779, piloted by 1st Lt. Carl Bostrom, 831st, took off at 0100 and 0120 hours respectively, to bomb the MUNICH WEST MARSHALLING YARD (PT) in Germany. The third aircraft, Q50, piloted by 2nd Lt. John H. Hazel, 828th, did not take off due to mechanical difficulties. The scheduled take off was delayed several minutes by local ground fog. The two aircraft arrived at Giovinazzo and departed on course at 11,100 feet at 0137 and 0143 hours respectively. No rendezvous with other Groups was effected and there was no fighter escort.

Aircraft 95517 returned prior to reaching the target at 0411 hours after feathering no. 3 engine. Aircraft encountered no flak or enemy aircraft. On return one searchlight was observed at Venice. The aircraft landed at base at 0633 hours. Inspection revealed that no. 3 engine had blown a cylinder.

Aircraft 48779 was over the target at 0508 hours and dropped 2 tons of 500 lb. RDX bombs by PFF from 23,000 feet. IIH flak was encountered over the target for 4 minutes, bursting high, low and wide to the left. As the bomb run began, the tail gunner reported that the searchlight beam flicked across the tail of the aircraft. There was an undercast over the target with some larger breaks in the target area, about 10 miles across. The 10/10 undercast at 4,000 feet cover the base. The aircraft landed at base at 0753 hours.

## UNIT HISTORY, 485TH BG (H) 1 DECEMBER 1944 TO 31 DECEMBER 1944

The 485th BG flew 14 operational mission during the month, making a grand total of 125 missions as of 31 December 1944. 668.75 tons of bombs were dropped during the month. Propaganda bombs were again used this month with delayed action bombs, with Blechhammer oil refinery in Germany as the target. Results on missions where bombing was done by visual means were generally good. The majority of the bombing during the month was done with the use of PFF equipment.

Five (5) aircraft and their crews were lost in combat during the month. One crew bailed out successfully over northern Italy and all members have returned to base. One crew bailed out near Rodi and. 4 members are still missing. Two crews bailed out over Yugoslavia and of these, all the members of one crew have been returned to base.

| | | | |
|---|---|---|---|
| Aircraft assigned 1 Dec. 1944 | 47 | Aircraft lost to salvage | 2 |
| Aircraft assigned 31 Dec. 1944 | 52 | Aircraft gained | 4 new PFF aircraft |
| Aircraft lost to combat | 4 | Aircraft gained from | 15 AFSC 7 |

Thirty-one (31) new crews arrived. For 12 days during the month, the weather conditions at the base were of such nature as to prevent flying. The mud never dried during the entire month. The weather, failing to provide a white Christmas, drove out the old year with the winter's first snow on December 31. Early in December, a 485th BG dance band, the Bombay Rhythm Lakers, was organized with personnel from the Group, 408th Service Sqdn and Co. A of the 1898 Engineers. The band made their official debut at Grant's Tomb on New Years Eve as part of the GI show - Welcome 1945. Instruments were procured through the Special Services Sect.

Members:

Sgt. Dudley A. Henriques - Leader
Captain William Wood - Drums
Lt. David Hichs - Piano
Lt. Charles Parham - Trombone
SSgt Edward Sykes - Trumpet
Pvt. Marshall Taylor - Trumpet
Pvt. Jimmy Cross - Trumpet

Sgt. Herbert Little - Saxophone
Sgt. Keith LeCheminant - Saxophone
Sgt. Jess Wood - Saxophone
TSgt. Herbert Wen - Guitar
Sgt. Robert Murphy - Base Violin
SSgt "Chuck" Panagos - Accordion

## MISSION NO. 112 - 3 DEC 1944

Two aircraft, 49029 and 48758, took off at 0736 and 0738 hours respectively, to bomb the primary target. But because observed weather conditions did not present sufficient cloud cover, the decision was made to bomb the KLAGENFURT MARSHALLING YARD in Austria. Aircraft 49029 was piloted by 2nd Lt. John H. Hazel, 828th and aircraft 48758 was piloted by Lt. Col. Griffin, 830th CO. Rendezvous with other Groups was not effected. Six to twelve P-51's were observed enroute and on the return in the vicinity of IP. There were no early returns. No enemy aircraft were seen although warning was received by radio, from other aircraft, that enemy aircraft were in the area.

Two aircraft were over the target at 1103 and 1152 hours, respectively, and dropped a total of 5 tons of 500 lb. RDX bombs by PFF. The target was completely cloud covered.

Weather: There were a few stratocumulus at 3500 feet over the base.

Aircraft 49029 and 48758 landed at 1305 end 1408 hours, respectively.

## MISSION NO. 113 - 6 DEC 1944

At 0709 hours 35 scheduled B-24's took off to bomb the primary target - HEGYESHALOM MARSHALLING YARD, HUNGARY. The 1st attack unit was led by major Harold A. Pruitt, 830 CO and the 2nd attack unit was led by Captain William D. Ceely, 831 Flt Cmdr.

Group assembly was accomplished at 0817 hours at 9,000 feet, which later assembled into Wing formation. Aircraft 29498 returned early.

Thirty-four aircraft were over the target at 1111 hours and 33 aircraft dropped 60 tons of 500 lb. GP bombs. Bombing was visual. Bombs fell in assigned area of the marshalling yard, causing several explosions among probable ammo trains.

No enemy fighters were seen and no flak encountered at the primary target. S-MIH flak was encountered enroute and return at Gyor for 3 minutes. Seven aircraft received minor flak damage. There were no casualties.

The weather was 3-4/10 stratocumulus at 7,000 feet and 7/10 at 15,000 feet over the base. 3/10 altostratus at 15,000 feet and visibility 20 miles.

Thirty-four aircraft landed at 1424 hours.

## MISSION NO. 114 - 8 DEC 1944

Two aircraft, 52034, piloted by 1st Lt. James W. Brady, 831st and 48779, piloted by 1st Lt. Richard L. Fedell, 830th, took off at 0155 and 0206 hours, respectively. The primary target was not bombed due to inadequate cloud protection in the area, and bombing of the 2nd alternate target was selected. Upon arrival there, no cloud cover was available over Graz, but a cloud layer existed to the east, hence, the bombs were dropped 15 miles east of Graz on a railroad junction at the town of Gliesdorf. There was no Wing formation and there was one early return. Aircraft 48779, turned back at 0222 hours because flame dampeners on no.1 and no. 4 superchargers were inoperative. The aircraft landed at base at 0358 hours.

One enemy aircraft was observed, showing one light, in the vicinity of Vodizze. The aircraft followed Aircraft 52034 on the right for 3-4 minutes, but did not come within firing range. No flak was encountered. Two rockets, which appeared like red flares, were observed below aircraft 52034 as the aircraft made landfall on the Yugoslavia coast, enroute to the target. One searchlight was seen at Graz but was not in motion. Two blinking lights were observed east of Graz, which appeared to follow the bomber. These were not search lights.

Aircraft 52034 was over the target at 0505 hours and dropped 2.5 tons of 500 lb. RDX bombs from 23,000 feet by PFF.

Weather: There were a few middle clouds over the base with visibility unlimited. There was a complete undercast over the target area.

Aircraft 52034 landed at base at 0746 hours.

## 10 December 1944

At 0813 hours, 29 B-24's took-off to bomb a primary target in Czechoslovakia. The aircraft were scheduled to assemble in three attack units. Lt. Col. Burton C. Andrus, Deputy Gp CO led the 1st attack unit; 1st Lt. Thomas H. McDowell, Jr., 828 Flt Cmdr led the 2nd attack unit and 1st Lt. Willie G. Geruson, 829 Pilot, led the 3rd attack unit. Assembly was accomplished over the home base. Some difficulty was experienced because of the existing low overcast and because aircraft from the 461 Gp were in the assembly area. The formation was recalled at 0855 hours before Wing rendezvous was effected because of unfavorable weather.

Weather: A high overcast with broken cloud layers and a ceiling of 1700 feet existed over the base at take-off. Visibility - 8 miles. Light rain began to fall at take-off, growing in intensity until the field was completely closed in at 1020 hours.

Aircraft 764 and 728 landed at base at 0908 and 0925 hours respectively. 27 aircraft landed at friendly fields:

| Lecce A/D | Bari A/D | Pantanella A/D | Gioia A/D | Brindisi A/D |
|---|---|---|---|---|
| 029 | 758 | 694 | 750 | 997 |
| 727 | 829 | 498 | | |
| 827 | 394 | 508 | | |
| 517 | 277 | | | |
| 038 | 908 | | | |
| 446 | 536 | | | |
| 779 | 504 | | | |
| 474 | | | | |
| 718 | | | | |
| 157 | | | | |
| 927 | | | | |
| 067 | | | | |
| 564 | | | | |
| 656 | | | | |
| 572 | | | | |

## MISSION NO. 115 - 12 DEC 1944

Four B-24's took off to bomb the BLECHHAMMER SOUTH OIL RIFINERY (PT) in Germany. Aircraft 758 - pilot 1st Lt. Joseph W. Cathey, 830th off at 0715 hours. Aircraft 779 - pilot 2nd Lt. Richard L. Fedell, 830th off at 0725 hours. Aircraft 517 - pilot. 1st Lt. Christian N. Schaefer, 831st - off at 0756 hours. Aircraft 410 - pilot 2nd Lt. Melvin H. Mooring, 829th - off at 0806 hours. There was no rendezvous or fighter escort.

Aircraft 758 was over the target at 1142 hours and dropped 2 tons of 500 lb. RDX bombs. IAH flak was encountered on the rally from the target for 6-7 minutes. The flak bursts were described as being black in color and always in groups of six.

Aircraft 779 lost turbo at the Yugoslav coast enroute and it was decided to bomb the 2nd alternate target - NORAVSKA/OSTRAVA OIL REFINERY in Czechoslovakia. The aircraft was over the target at 1113 hours end dropped 2 tons of 500 lb. RDX bombs and a propaganda bomb. SIH flak was experienced over the target for 2 minutes. The flak was bursting at the same level as the aircraft but, behind and to either side. Aircraft 779 reported seeing 5 unidentified single-engine aircraft which made no attempt to attack. Bombing was by PFF.

Aircraft 517 and 410 returned, early.

Aircraft 758 returned to base at 1602 hours and aircraft 779 returned at 1556 hours. Weather: There was 8-9/10s stratocumulus at 2500 feet. The target was covered by an undercast.

## MISSION NO. 116 - 15 DEC 1944

At 0814 hours, 28 B-24's, scheduled, and 1 spare aircraft took off to bomb the SALZBURG MARSHALLING YARD (PT) in Austria. The formation was led by Lt. Col. Andrus, Deputy Gp CO. Difficulty was experienced in the Group assembly because of the area being overcast - 10/10 stratocumulus clouds - with tops at 7500 feet. Twenty-one (21) aircraft assembled over the field at 0830 hours at 8,000 feet. Six aircraft did not assemble as they were unable to locate the formation. The spare aircraft and one other aircraft returned to base prior to assembly. The Wing leader requested UPHILL to extend the rendezvous time by five minutes. A Wing line rendezvous was effected at 0857 hours. Fifteen (15) - twenty (20) P-51's escorted the formation from 1008 to 1236 hours. Aircraft 908, 474, 495, 067, 718 and 717 returned early.

Twenty-one (21) aircraft were over the target at 1137 hours and dropped 31.5 tons of 500 lb. RDX bombs and 5 propaganda bombs by PFF. The results were undetermined because of a complete undercast. MIH to MAH flak was reported over the target for 3 minutes, bursting behind and to one side in the vicinity of the dispersed chaff. One crew reported one - four gun battery slightly east of Mandeel. No enemy aircraft were seen. Two aircraft received minor flak damage and SSgt Robert L. Iman received a minor flak wound.

Weather: There was a stratocumulus overcast at 6500 feet at take off and 6-8/10s at 2500 on return.

Twenty-two (22) aircraft landed at base at 1436 hours.

## MISSION NO. 117 - 16 DEC 1944

At 0843 hours, 33 aircraft took off to bomb a primary target in Germany. Lt. Col. Nett, 828th CO led the formation. The Group assembled into Wing formation. Twenty (20) P-38's escorted the formation from 1220 to 1433 hours. Aircraft 495, 827, 758, 572, 517, 526 and 997 returned early.

Enroute, the PFF equipment in the lead aircraft became inoperative. No other PFF aircraft being present, the Group leader determined to follow the 460th BG over the target and bomb with them. The IP was reached and a bomb run was begun on the target. For some reason, the 460th abandoned this target. The 485th, having no alternative, followed the 460th, returning to the 1st alternate target.

Twenty-six (26) aircraft were Pilsen, Czechoslovakia at 1315 hours at 24,500 feet and 25 aircraft dropped 36 tons of 500 lb. RDX bombs by PFF of the 460th BG. No enemy aircraft were encountered. One unidentified aircraft and 1 Me 210 were seen near Lichtenstadt. Neither aircraft made an attempt to attack the formation. MIH flak was encountered at Pilsen for 4 minutes. Three aircraft received minor flak damage. A vertical wall of flak was observed at the primary target.

Weather: It was clear over the base.

Aircraft 508 landed at Nature airdrome; aircraft 227 landed at Guilia airdrome and aircraft 699 landed at Lesi airdrome.

Twenty-three (23) aircraft landed at base at 1716 hours.

831st Bomb Sqdn Aircraft Blue "S"

| | | |
|---|---|---|
| P | 2nd Lt. Richard D. Kingsbury | Reportedly, Blue "S" left the formation |
| CP | 2nd Lt. David T. Hansen | at north end of the Adriatic Sea on return |
| N | F/O Anthony F. Treharns | and headed for the coast. Believed |
| CP | 2nd Lt. Harley Beard | intended to land at friendly field for |
| ENG | Sgt. Frank Camichle | fuel. Aircraft, apparently under control. |
| RO | Sgt. William B. Highbe | |
| TG | Sgt. Burton E. Van Dellon | |
| UG | Sgt. Richard T. Gardner | |
| NG | Sgt. Randolph E. Russell | |
| AG | Sgt. Robert G. Evans | |

## MISSION NO. 118 - 17 DEC 1944

At 0809 hours, 26 of 29 scheduled B-24's took off to bomb the primary target - SOUTH BLECHHAMMER SYNTHETIC OIL REFINERY. The 1st attack unit was led by Lt. Col. John E. Atkinson, 831 CO and the 2nd attack unit was led by Capt. Eugene B. Lenfest, 830 Flt Cmdr.

The Group assembled into Wing formation. There were no early returns. Twenty-six aircraft were over the target at 1234 hours and 24 aircraft dropped 48 tons of 500 lb. RDX's and one propaganda bomb from 24,500 feet. Bombing was done by PFF.

No enemy aircraft were seen. MIH flak was encountered in the target area for 3 minutes. MI-SAH flak near Gyor and SAH flak at Nagyhanizsa was encountered, the latter two areas for

1 to 2 minutes. Rockets fired from the ground were also observed. Six aircraft received minor flak damage. There were no casualties.

The weather over the base was overcast at 6,000 feet and the target area was overcast up to 12,000 feet.

Twenty-six aircraft landed at 1622 hours.

### MISSION NO. 119 - 18 DEC 1944

At 0749 hours, 28 aircraft took off to bomb the BLECHHAMMER SOUTH OIL REFINERY (PT) in Germany. The 1st attack unit was led by Lt. Col. Reeve, 829th CO and the 2nd attack unit was led by 1st Lt. William R. Fritz, 828th Flt Cmdr. Wing rendezvous was effected. Sixty (60 to eighty (80) P-38's and P-51's escorted the formation from 0951 to 1350 hours. Aircraft 474 and 157 returned early.

Enroute, three aircraft of Charlie box fell behind the formation. Being unable to rejoin the formation, they continued on course and shortly dropped 6 tons of 500 lb. GP bombs from 23,000 feet at 1135 hours. On return, these aircraft encountered SIH flak over Gyor. At Zagreb, Yugoslavia, IAH flak was encountered at 16,000 feet for 4-minutes. Aircraft 51872 was critically damaged. With no. 2 engine feathered and with no. 3 engine pouring a stream of fuel, the aircraft went into a gradual dive at which time the crew started bailing out. After the last man had jumped, ten in all, the aircraft went into a steep dive through the undercast. Aircraft 495 and 335 received minor flak damage. Twenty-three (23) aircraft were over the target at 1207 hours and 22 aircraft dropped 44 tons of 500 lb. GP bombs from 24,000 feet by PFF. Results of the bombing were undetermined as the target was completely cloud covered.

No enemy aircraft were seen. Flak encountered at the primary target was SIH, increasing with intensity to the left of the formation. However, no aircraft were damaged.

In addition to aircraft 51827 lost to flak, aircraft 78508 was last seen over the spur of Italy on return. It was later reported to have ditched near Rodi. Seven crewmembers have been accounted for and 4 members are missing.

Weather: There was an overcast at the base and an undercast at 14,000 feet over the target area. Twenty-four aircraft larded at base at 1533 hours.

828th Bomb Sqdn Aircraft 50827

| | |
|---|---|
| P | Captain Joseph S. Gill |
| CP | 1st Lt. Asimakes S. Maniatty |
| N | 2nd Lt. Billy T. Hutcheson |
| B | 2nd Lt. Albert A. Frinche |
| ENG | Sgt. John Patucek |
| RO | MSgt. Charles J. Rogers |
| NG | Sgt. Richard F. Mitchell |
| UG | Sgt. Paul L. Alexander |
| BG | Sgt. Gottfried Helwer |
| TG | Sgt. Harold J. Weaver |

Aircraft was hit by flank at Zagreb at 1255 hours. No. 2 engine feathered, No. 3 streaming fuel. Plane went into a gradual dive. Ten chutes were seen to opened.

831st Bomb Sqdn Aircraft 78508

| | |
|---|---|
| P | Captain John C. Buker |
| CP | 1st Lt. Robert (nmi) Gillette |
| CP | 2nd Lt. Harley E. Beard |
| N | 2nd Lt. Jacob (nmi) Bloomfield |
| N | 2nd Lt. Willis G. Bloomquist |
| B | 2nd Lt. Harry (nmi) Twichell |
| ENG | Sgt. John F. Driscoll |
| RO | Sgt. Paul M. Douglas |
| TG | Sgt. Edwin O. Wright, Jr. |
| UG | SSgt. John J. Hirsch, Jr. |
| BG | Sgt. James H. Favre |

Reportedly, that aircraft dropped its wheels in the vicinity of the spur on the return route as if going to land at a friendly field.

## MISSION NO. 120 - 19 DEC 1944

At 0807 hours, 26 B-24's took off to bomb a primary target in Germany. The 1st attack unit was led by Major Harold A. Pruitt, 830th CC and the 2nd attack unit led by Captain Gerald H. Morris, Jr., 831st Flt. Cmdr. The Group assembled into Wing formation. Thirty (30) P-38's escorted the formation from 1045 to 1300 hours. Several P-51's joined the formation near Lipnik.

Enroute, over the Adriatic, the 485th BG intercepted the 47th Bomb Wing, which was heading northwest at the same altitude. In order to avoid collision with the rear units of the 47th BW, this Group veered to the left. The maneuvering necessitated reassembling and the Group was back on course over Split 15 minutes later. Aircraft 486 and 699 returned early.

Adverse weather was encountered the undercast and overcast converged in that vicinity, resulting in conditions making it impossible to make a bomb run on PFF. Decision was made to bomb the 2nd alternate target Maribor, Yugoslavia. Again, the formation became dispersed and several of the aircraft evidently, confused, jettisoned their bombs. Maribor was bombed by PFF. The bombs failed to release from the lead aircraft and several of the aircraft of the 1st unit held their bombs. The 2nd unit got their bombs away. Nine (9) aircraft dropped 18 tons of 500 lb. RDX bombs from 23,000 feet at 1314 hours. No enemy aircraft were seen. SIH flak was encountered over Maribor for 2 minutes. One aircraft received minor flak damage.

Aircraft 656 left the formation, being unable to keep up with it and landed at home base slightly ahead of the formation.

Aircraft 997 crashed in a valley on the Isle of Vis. Reported 2 crewmembers were killed.
Aircraft 504 landed at Madna airdrome.
Weather: There was a 7-9/10 overcast at 7,000 feet.
Twenty-two (22) aircraft landed at base at 1520 hours.

## MISSION NO. 121 - 20 DEC 1944

Twenty-five (25) B-24's (27 scheduled) took off to bomb the BRUX SYNTHETIC OIL REFINERY in Czechoslovakia. The 1st attack unit was led by Lt. Col. Nett, 828th CO and the 2nd attack unit was led by Captain Lenfest, 830th Flt Cmdr. Wing formation was effected and 25 P-38's escorted the formation from 1115 to 1420 hours. Aircraft 460, 468, 829, 276 and 801 returned early.

Twenty (20) aircraft were over the target at 1248 hours and dropped 40 tons of 500 lb. RDX bombs from 25,000 feet of PFF. The results were undetermined as the target was cloud covered. No enemy aircraft were encountered. One Me 109 was seen. IAH flak encountered over the target for 5 minutes.

Weather: There was an overcast over the base at 2500 feet.

Eighteen (18) aircraft were over the field at 1609 hours. Aircraft 474 landed at 1615 hours with a flat nose wheel and blocked the runway for several minutes. Eight aircraft landed at 1627 hours. Clouds moved in suddenly over the field causing it to be closed for further landing. Ten aircraft landed at Nature field, one aircraft landed at Amendola field and one aircraft is missing - 51277, which is reported to have crashed over the spur of Italy on return. Crew reported to have bailed out.

## MISSION NO. 122 - 26 DEC 1944

At 0815 hours, 25 B-24's (27 scheduled) took off to bomb the BLECHHAMMER SOUTH OIL REFINERY (PT) in Germany. The 1st attack unit was led by Lt. Col. Douglas M. Cairns, Gp Ops Officer and the second attack unit was led by Captain Gerald H. Morris, Jr., 831st Flt Cmdr. Wing formation was effected and 30 to 50 P-38's escorted the formation, with several P-51's joining at 1300 hours. The P-38's departed the formation at 1214 hours. Aircraft 498 and 426 returned early.

Twenty-three aircraft were over the target at 1230 hours and dropped 45.5 tons of 500 lb. RDX bombs from 25,300 feet, visually. Results: Photo showed the target to be almost completely smoke covered. Few strikes in the central portion of the area and a few are discernible in the north section near the briefed MPI. No enemy aircraft were seen. IAH flak was encountered for 4 minutes. Aircraft 50486 received a direct hit over the target, which completely severed the tail assembly. Aircraft went into a dive and continued in the dive until out of sight. It is believed that one crew member jumped before plane went into the dive as one chute was reported as opening in that vicinity. Six (6) aircraft received minor flak damage.

Aircraft 29494 crashed at 1427 hours. The pilot called On radio, stating that the crew was going to bail out is the aircraft was short of fuel. Nine chutes were seen to open. Aircraft 28834 and 51335 landed at Vis for fuel and have not returned to base at this time.

Weather: It was clear over the base, and over the target area.

Nineteen (10) aircraft landed at base at 1553 hours.

831st Bomb Sqdn Aircraft 29494

| | | | |
|---|---|---|---|
| P | 1st Lt. David A. Blood | G | SSgt. Robert F. Lolvet |
| CP | 1st Lt. Lewis B. Baker | G | SSgt. Thomas T. Tamraz |
| B | F/O Eugene D. Cogburn | G | Sgt. Wilmont M. Gibson |

| | | |
|---|---|---|
| N | F/O George E. Benedict | FLT SURGEON Captain James B. Johnson |
| ENG | SSgt. Fred B. Sherer | |
| RO | TSgt. Michael Yaworsky | Aircraft crashed about seven miles north |
| G | Sgt. Warren A. La France | east of Frval and Frekoga.  Nine chutes were seen to opened. |

828th Bomb Sqdn Aircraft 50486

P  1st Lt. Arthur E. Lindell
CP  2nd Lt. Avery Gilliland
N  2nd Lt. Howard (nmi) Boslow
ENG  Sgt. Alex Abramowich
RO  Sgt. Joseph D. Ryan
G  SSgt. Joseph F. Lajkowicz
G  Sgt. Travis E. Burns
G  Sgt. Myron L. Yaw
G  Sgt. Michael C. Papadopulos

Aircraft received a direct flak hit over the target and went down in a dive.  The tail section broke off.  One chute seen at 1230 hours.  Tail section reported seen blown of aircraft.  One crewmember watched it go down for at least 15,000 feet and no chutes were observed.

## MISSION NO. 123 - 27 DEC 1944

At 0916 hours, 27 of 28 scheduled B-24's took off to bomb the primary target - MARIBOR MARSHALLING YARD.  The 1st attack unit was led by Lt. Col. Andrus, Deputy Gp CO and the 2nd attack unit was led by Lt. Col. Nett, 828 CO.

The Group rendezvous with the 450 Gp and proceeded to target without fighter escort.  There were no early returns.

Twenty-seven aircraft were over the target at 1232 hours and 19 aircraft dropped 47.5 tons of 500 lb. GP's from 21,400 feet, visually.  Target was almost completely smoke covered from previous bombing, making it impossible for the boxes to aim on briefed MPI's.  After rallying to the right to Dittmansdorf, the formation proceeded to KALGENFURT MARSHALLING YARD, where 2 aircraft dropped 5 tons from 21,500 at 1248 hours visually.  The formation then proceeded to VILLACH and made a run over the NORTH MARSHALLING YARDS, where 5 aircraft dropped 12.5 tons at 1303 hours from 21,500 feet, visually.

Bombing results:  At Maribor, hits were made on the north choke point; at Klagenfurt the target was missed and at Villach bombs fell at the west end of the north marshalling yard, damaging the station, sidings, warehouse and cutting of the tracks.

No enemy aircraft were encountered.  Between Klagenfurt and Villach, 2 Me 262's were observed 6,000 feet above the formation, heading south.  They turned and circled the formation 3 times without making a pass.  One crew reported seeing the same 2 aircraft give chase to a P-38 in the vicinity of Villach.

MAH flak was encountered over Maribor.  No flak was encountered over Klagenfurt or Villach.  Four aircraft received minor flak damage and one aircraft received major flak damage.

There was one casualty - Sgt. Milto Wolfson, Gunner, received a slight injury in the leg from flak.

The weather was clear over the base and in the target area.

Twenty-seven aircraft landed at 1545 hours.

## MISSION NO. 124 - 28 DEC 1944

At 0828 hours, 24 B-24's (26 scheduled) took off to bomb the KRALUPY OIL REFINERY in Czechoslovakia. The 1st attack unit was led by Lt. Col. Nett, 828th CO and the 2nd attack unit was led by Captain Lloyd Allan, 829th Flt Cmdr. The Group assembled into Wing formation. At 1125 hours, 30 P-38's rendezvoued with the formation and at 1141 hours, joined by 35 P-51's. Several more P-51's joined in escorting the formation until 1358 hours. Aircraft 725 had a runway prop on no. 1 engine and returned early.

Twenty-three (23) aircraft were over the target at 1222 hours and dropped 43 tons of 500 lb. RDX bombs from 22,000 feet by PFF. The results were undetermined. No enemy aircraft were seen, no flak was encountered and there were no casualties.

Weather: It was clear over the base.

Twenty-three (23) aircraft landed landed at 1626 hours.

## MISSION NO. 125 - 29 DEC 1944

At 0911 hours, 27 B-24's, scheduled, took off to bomb the VERONA PORTO VESCONA MARSHALLING YARD in Northern Italy. The 1st attack unit was led by Major Harold A. Pruitt, 830th CO and the 2nd attack unit was led by 1st Lt. William R. Fritz, 831st Flt Cmdr. The Group assembled into Wing formation. There was no rendezvous with fighter escort. Several P-38's and P-51's were reported circling in the general target area. There were no early returns.

Twenty-six (26) aircraft were over the target at 1245 hours and dropped 64 tons of 500 lb. bombs visually. Ten strikes were in the yard near the MPI. Thirty to forty wagons were destroyed or damaged. One hundred strikes were south of the yard damaging several small buildings. Five strikes were on double railroad tracks southeast of the marshalling yard.

Several strikes were in the engine iron works southwest of the yard and 10 strikes were in the yard and 10 strikes were in the barracks and stores west of the yard.

No enemy aircraft were seen. MIH to MAH flak was encountered over the target for 5 minutes. The flak burst lagged the formation with only a few burst being accurate. Aircraft returned to base were not damaged. Aircraft 424 landed at Foggia with wounded crewmember.

Since the weather at the base was very uncertain, the group CO flew over the Adriatic in a B-25 to meet the formation and to guide it to base. He located the formation flying very low and steered it around the spur and started to base. In the vicinity of Cerignola, a layer of haze was encountered causing three of the Boxes to turn back to friendly airfields.

Weather: On take off there was 5/10 overcast at 2000 feet. On return there was a complete overcast and light rain with visibility of 2 miles.

Six aircraft landed at base at 1535 hours. Seventeen aircraft landed at Ramitelli. One aircraft landed at Foggia, one at Bari, and one at Lesina. Aircraft 10572 crash-landed at Cutella airdrome. The aircraft was completely wrecked and no personal were injured.

# UNIT HISTORY, 485TH BG (H) 1 JANUARY 1945 TO 31 JANUARY 1945

During the month of January, the 485th BG flew 4 operational missions, making a grand total of 129 missions as of 31 January 1945. 103.7 tons of bombs were dropped.

Four (4) aircraft and their crews were lost during the month. One crew bailed out over Yugoslavia and all 10 men returned to base later. The other three aircraft were ditched in the Adriatic Sea with 4 known survivors from these three aircraft.

| | |
|---|---|
| Aircraft on hand - 1 Jan 1945 | 52 |
| Aircraft on hand - 31 Jan 1945 | 55 |
| Aircraft lost to operational accidents | 2 |
| Aircraft lost to non-operational accidents | 2 |
| Aircraft lost to combat | 4 |
| Aircraft transferred to 15 AF Service Command | |
| Aircraft gained (new) | |
| Aircraft repaired by 15th AF service Command | 5 |

AWARDS

| | |
|---|---|
| DFC | 12 |
| SILVER STAR | 1 |
| PURPLE HEART | 13 |

Combat inactivity caused some unrest and dissatisfaction among the personnel. There were numerous briefings with crews getting up early, attending briefings, commuting to the line, only to have the mission cancelled at the last minute prior to taxiing out for take off was a regular occurrence.

## MISSION NO. 126 - 8 JAN 1945

At 0809 hours, 25 of 28 scheduled B-24's took off to bomb the primary target in Austria. The 1st attack unit was led by Lt. Col. Roy L. Reeve, 829 CO and the 2nd attack unit was led by 1st Lt. Thomas H. McDowell, 828 Flt Cmdr.

The Group assembled over the home base at 3300 feet at 0823 hours. Aircraft 95460 crashed on the field during assembly. Six crewmembers parachuted safely; the remaining 4 crewmembers were killed in the crash.

The Group assembled into Wing, formation and 50 P-38's provided fighter escort from 1115 to 1315 hours. Aircraft 834 returned early - turning back over the Adriatic after turbos malfunctioned. The formation departed on course from Spinazzola and remained until reaching the Adriatic where on a course to the left of the briefed course was followed because of weather concentrated along the Yugoslavian coast. Upon reaching the IP, the Wing formation veered to the right. The 485th followed the Wing leader until 1130 hours, when the Wing leader turned to the right into a cloud layer - the 485th avoided the clouds by turning left. At 1155 hours, the Wing leader advised all Groups by radio to bomb the AF No. 3 target. A bomb run was made on a target which evidence indicates was located somewhere in the Salzburg area. "A" and "B" boxes dropped on the target - 8 aircraft dropped 16 tons of 500 lb. RDX's from 25,500 feet.

"C" box became separated from the formation at Caorle due to the existing cloudbanks through which the formation was flying. Circling around Bruck, they continued to Salzburg, changing course to Villach where the Marshalling Yard was bombed by PFF - 7 aircraft dropped 12.5 tons of 500 lb. RDX's at 1245 hours from 25,700 feet. After leaving Caorle, "D" box lost the formation. There being no PFF aircraft in this box and due to an undercast existing in the area, the leader decided to return to base, turning back at Spittal. All targets were cloud obscured and bombing was by PFF.

No enemy aircraft were seen. SAH flak was encountered in the Salzburg area for 1 minute. No damage here. SAH flak was encountered at Villach for 2 minutes.

Aircraft 038 received minor flak damaged in this area.

The weather over the base was 5 to 7/10 at 3,000 feet. Twenty-three aircraft landed at 1525 hours.

## MISSION NO. 127 - 19 JAN 1945

At 0827 hours, 28 B-24's took off to bomb the primary target - ZAGREB EAST MARSHALLING YARD, YUGOSLAVIA. The 1st attack unit was led by Lt. Col. John E. Atkinson, 831st CO and the 2nd attack unit was led by Captain Charles E. Porter, 830th Asst. Ops Officer.

The Group did not rendezvous with the Wing formation or fighter escort. Aircraft 727 returned early due to a runaway prop.

At 0930 hours, Col. Atkinson, after experiencing a complete radio failure of the lead aircraft, relinquished the lead to Captain Gerald H. Morris, 831st Flt Cmdr. Over Yugoslavia, a complete undercast at 13,000 feet was encountered. The leader, after determining that no targets of plan Baker could be bombed visually and after receiving a message by radio to

bomb no targets of plan Able, decided to return to base. The formation became slightly scattered at the point of turn-back with several of the aircraft temporally losing the formation.

Aircraft 764 lost the formation and flew alone for several minutes, during which time the primary target was identified through a break in the undercast, dropped 2 tons of 500 lb. GP's from 19,000 feet at 1208 hours. Results were undetermined.

No enemy aircraft were encountered. Several unidentified aircraft were seen. One was believed to be a HE 111 and 5 single-engine aircraft, probably ME 109's, were also seen. No flak was encountered.

The weather over the base was clear. Aircraft 725 cracked up on landing, delaying the 5 aircraft from landing for several minutes. Twenty-seven aircraft landed at 1426 hours.

## MISSION NO. 128 - 20 JAN 1945

At 0824 hours, 28 B-24's took off to bomb the LINZ COMMUNICATIONS INSTALLATIONS in Austria. The 1st attack unit was led by Col. John P. Tomhave, Gp CO and the 2nd unit was led by Major Phillips E. Cummings, 829th Ops Officer. Wing formation was effected. Thirty (30) P-38's rendezvous with the formation at 1145 hours and at 1250 hours, 25 P-51's were seen in the target area. Aircraft 628 and 038 returned early.

Enroute, at approximately 30 miles off the Italian coast, the weather began to build up. This condition forced the formation to climb to altitude sooner than had been briefed. However, after crossing the Alps, the formation broke through the weather and from there to the target favorable weather existed. The bomb run was visual.

Twenty-six (26) aircraft were over the target at 1254 hours and dropped 45.2 tons of 100 lb. clustered bombs. Due to the photo coverage obtained, the results are for the most part undetermined. Photos from the lead box indicated approximately 20 strikes in the general target area. The snow covered the terrain, made the target identification difficult and also hindered the spotting of an accurate bomb plot. No enemy aircraft were seen. IAH flak was encountered over the target. MH flak was observed at Steyr. Sixteen (16) aircraft incurred minor flak damage. Five aircraft were not damaged at all.

Two aircraft landed at friendly fields and three aircraft ditched. Aircraft 718 ditched at 43-07N & 15-34E at 1503 hours. Aircraft 750 ditched at 42-00N & 15-43E at 1523 hours. Aircraft 699 ditched near the Isle of Vis. Aircraft 770 landed at Ancona, refueled and returned later. Aircraft 495 crash-landed at Madna and aircraft 067 landed at Vis for fuel.

Weather: It was clear over the base at take off with 1/10 stratocumulus at 3000 feet on return.

Twenty (20) aircraft landed at base at 1554.

## MISSION NO. 129 - 31 JAN 1945

At 0841 hours, 15 of 18-scheduled B-24's, known as Red Force, took off to bomb the primary target - MOOSBIERBAUM OIL PEFINERY in Austria. Wing formation was effected and 30 P-51's escorted the Group from 1200 to 1419 hours.

Fourteen aircraft were over the target at 1325 hours and dropped 28 tons of 500 lb. GP's from 26,000 feet by PFF, as target was completely overcast.

No enemy aircraft were seen. Flak at the target was SIH, bursting below and behind the formation. SIH flak, probably railroad flak, encountered east of Gyor. No aircraft were damaged.

Thirteen aircraft landed at 1545 hours. Aircraft 41157 landed at Lesing, refueled and returned to base.

At 0941 hours, 15 of 17-scheduled B-24's, known as Blue Force, took off to bomb the same primary target as Red Force. Blue Force was led by Lt. Col. Andrus, Deputy Gp CO. Fourteen aircraft assembled, with the last aircraft taking off, failed to find the formation and joined and bombed with the 304th Wing formation. Wing formation was effected and 30 P-51's escorted the Group from. 1249 to 1500 hours. Aircraft 394 returned early.

Thirteen aircraft were over the target at 1436 hours. Ten aircraft, including the one with the 304th Wing dropped 20 tons of lb. GP's. "C" box did not dropped its bombs at this time. Three aircraft of this box dropped 6 tons of bombs on the Maribor Marshalling Yard at 1504 hours from 24,300 feet.
Results: Undetermined.

There was a thin overcast at 9,000 feet over the home base. Eight aircraft returned at 1656 hours. Five aircraft landed at Vis for gas and have not returned to base: 410, 446, 335, 536 and 498. 51992 is missing was last observed as the formation left the target. It was in no apparent trouble.

## UNIT HISTORY, 485TH BG (H) 1 FEBRUARY 1945 TO 28 FEBRUARY 1945

During the month of February, the 485th BG flew 20 operational missions, making a grand total of 149 missions as of 28 February 1945. 968.55 tons of bombs were dropped on targets, mainly by visual means, a new experience for many of the bombardiers, since most of the missions of the previous 2 months had been made by PFF.

On these missions, 5 aircraft and their crews were lost from enemy action. However, one of these crews was later, evacuated from, Yugoslavia. Also during the month, 9 men from a crew reported MIA on January 31, 1945 returned to base.

There was one other combat casualty during the month. The navigator of aircraft 882, hit by flak over Pola, was killed and blowing free by the burst which tore a large hole in the nose section of the aircraft.

| | |
|---|---|
| Aircraft on hand - 1 Feb 1945 | 52 + 2 ww |
| Aircraft on hand 28 Feb 1945 | 60 + 2 ww |
| Aircraft lost to operational accidents | 7 |
| Aircraft lost to non-operational accidents | 0 |
| Aircraft lost to combat | 6 |
| Aircraft transferred to 15th AF Service Command | 3 |

AWARDS

| | |
|---|---|
| DFC | 11 |
| SOLDIERS MEDAL | 2 |
| BRONZE STAR | 2 |
| PURPLE HEARTS | 29 |

## MISSION NO. 130 - 1 FEB 1945

At 0832 hours, 25 B-24's (26 scheduled) took off to bomb a primary target. Lt. Col. Nett, 828th CO led the 1st attack unit and Captain William D. Ceely, 831st Ops Officer led the 2nd attack unit. Fifty (50) P-51's escorted the Group formation from 0930 to 1405 hours. The Group formed with the 460th BG and Wing formation was effected. Aircraft 727 returned early.

The formation flew on the briefed course, reaching Kutina at 1125 hours, where instructions were received by the Group leader from the Wing leader to bomb the 2nd alternate target - AF target no. 2. A bomb run was started by PFF. However on the bomb run, another Group suddenly changed their course which placed the 485th BG in the other Group prop wash. A 360-degree turn was made and a 2nd bomb run was started. On this run the target could not be found in the PFF scope. The Group leader then decided to bomb the 1st alternate target - AF target no. 1. After reaching a point approximately 40 miles east of this target, it was determined that a visual run couldn't be made. The formation returned to base.

In the Graz, Austria area, "D" box leader saw chaff being dispensed and immediately called to find out if the formation on the bomb run. The reply was misunderstood by the bombardier who released his bombs following by the remaining six aircraft, in the box. Interrogation failed to disclose the exact area of bombing. Photos showed a bomb pattern in a flat open country with scattered trees. Bombs were dropped at 1221 hours from 23,500 feet. Seven (7) aircraft dropped 13.75 tons of bombs in the Graz area.

No enemy aircraft were seen. SIH flak was encountered at Zagreb and Graz. Flak was the predicted type with bursts mainly off course and to the rear. Aircraft 458 received minor flak damage.

Weather: it was clear over the base at take off and 6/10s coverage at 3500 feet on return. There was a 9 to 10/10 cloud coverage over the target area.

Twenty-four (24) aircraft landed at base at 1527 hours.

CHAFF - Strips of tin foil were dispensed by the waist gunners to produce a false signal to confuse the radar controlled anti-aircraft guns.

## MISSION NO. 131 - 5 FEB 1945

At 0820 hours, 37 of 38 scheduled B-24's took off to bomb the REGENSBURG OIL STORAGE facilities in Germany. Lt. Col. Reeve, 829 CO led the formation with three attack units of two boxes each.

The Group assembled into Wing formation. Twenty P-51's escorted the Group from 1155 to 1417 hours. Aircraft 546, 394, 859 and 034 returned early.

The IP and target area were partially cloud covered with a layer of stratocumulus at 8,000 feet. The clouds drifted from over the target, permitting the Bombardier to see the aiming point. In his haste to change to a visual bomb run, the Bombardier accidentally hit the release switch, resulting in the 1st attack unit's bombs falling approximately 3 miles short of the target.

The other 2 attack units, after being notified by the formation leader that the lead unit had dropped early continued their bomb run on PFF and released on target.

Thirty-two aircraft were over the target at 1302 hours and dropped 63.5 tons of 250 lb. GP's from 23,000 feet. Bomb strike photos showed the target area to be approximately 8/10 cloud, covered. Thirty-five bomb bursts are visible approximately 1 mile south of the target.

The weather over the base was good with scattered altocumulus at 9,000 feet with a few high clouds on return.

No enemy aircraft were encountered. One crew reported seeing an Me110 in the target area. The formation heard over the radio that another Group had been attacked by enemy aircraft. The escorting aircraft were milling over the target area, indicating that enemy aircraft were in the area. Flak at the target was MIH, and of the predicated type for 3 minutes. Aircraft 656 received minor flak damage.

Thirty-three aircraft landed at 1618 hours. A bomb accidentally released from the fifth aircraft to land, blocked the runway for several minutes, delaying the remaining aircraft from landing until it was removed.

**MISSION 131, 5 FEBRUARY 1945. TARGET: REGENSBURG OIL STORAGE FACILITY, GERMANY. BOMBS: 250 LB GP, BOMBING ALTITUDE: 23,000 FEET.**

The Messerschmitt final assembly plant was located here. Since they were turning out 200-300 aircraft each month, fuel was needed to fly them to fighter bases. Although the target shows the target totally covered, the clouds drifted from over the target long enough for the lead bombardier to see the aiming point.

**MISSION 132, 7 FEBRUARY 1945.  Target:  POLA OIL STORAGE FACILITIES.**

**On this mission Aircraft 44-49882 (Blue L) from the 831st Squadron received a direct hit in the nose, killing Lt. Donald Swenson and trapping the nose gunner, Robert Espenshade, in his turret. Espenshade, seen here standing in the nose, received only minor injuries.**

## MISSION NO. 132 - 7 FEB 1945

Forty B-24's took off to bomb the POLA OIL STORAGE FACILITIES in Italy. The 1st attack unit was led by Lt. Col. john E. Atkinson, 831st CO and the 2nd attack unit was led by Major Harold E. Pruitt, 830th CO. The Group did not rendezvous with other Groups or fighter escort. There were no early returns.

Forty aircraft were over the target at 1412 hours and dropped 76.5 tons of 500 RDX's from 20,000 feet. Bombing was visual. Photos revealed approximately 50 strikes in the target area, causing several fires along the storage tanks.

There was an overcast at 2500 feet over the base with lower broken clouds at 800 feet and visibility was 4 miles. The target area was clear.

MAH flak vas encountered over the target for 2 minutes. The flak started below and behind the first unit; rapidly found the correct range and altitude. Burst came from shore emplacements, appearing three at a time. Eleven aircraft of the first force received flak damage: 882 received major damaged and was salvaged; 410 received several holes in the nose section end in both wings; - requires 7 days to repair; 727 was hit in the left tokyo tank, no. 2 oil tank and control cables in the bomb bay section requiring 6 days to repair; 596 received over 100 holes, damaging engines, props, rudder, gas tanks and flaps - requiring 5 days to repair. 564 had a wing tank damaged and hits in no. 4 engine which had to be feathered - requires 3 days to repair; 438 received a flak hole through a cylinder of no. 1 engine and holes in the right navigator window, requiring 2 days to repair and 276 received hits in the elevator, wings, bomb bay section and vertical stabilizers - requires 2 days to repair. 638 received several small flak holes requiring 1 day to repair; 899 received holes in the tail turret and 592 received several small holes in the fuselage. There were 3 casualties.

882 received a direct hit in the nose section, killing the navigator and throwing him from the ship. The nose gunner was pinned in the nose turret and was released after the aircraft had landed. He suffered minor bruises and injuries.

Aircraft 834 landed at Lucerna with a wounded navigator.

The 2nd unit, bombing from approximately 2,000 feet higher than the 1st unit encountered only scant, inaccurate flak. No aircraft of this unit received flak damage. Thirty-eight aircraft landed between 1543 and 1624 hours.

## MISSION NO. 133 - 8 FEB 1944

At 0825 hours, 28 B-24's took off to bomb the VIENNA SOUTH GOODS DEPOT in Austria. The 1st attack unit was led by Lt. Col. Andrus, Deputy Gp CO and the 2nd attack unit was led by 1st Lt. William G. Ferguson, 829 Flt Cmdr.

Wing rendezvous was accomplished and 36 P-51's escorted the Group from 1220 to 1340 hours. Aircraft 930 returned early because the plane could not keep up with the formation, lost sight of the formation at 1235 hours and landed at base at 1446 hours. Aircraft 812 had inverter trouble and no 2 turbo ran away, causing the aircraft to return early.

In the vicinity of the Spur, the wing formation received instructions from Wing HQ, to deviate from the briefed course. Later, a box from the 449th Gp, led by aircraft 78, interfered with this Gp, causing a near collision. However, the IP was located and the bomb run made on PFF.

Twenty-six aircraft were over the target at 1257 hours and dropped 52 tons from 26,000 feet. Photos revealed the target to be approximately 10/10 cloud covered. It is believed that the bombs fell within the city and in the target area. No enemy aircraft were encountered. Eight to twelve enemy aircraft were reported in the area at 47 - 15N and 15 - 08E. These aircraft were flying in formation and were identified as Me 262's and FW 190's. Either a mid-air collision or a dog fight was seen in or adjacent to this formation which resulted in two aircraft going down. One parachute was observed.

Flak at the target was MIH, partly barrage and partly predicted type, and was mostly below the formation with 8 - 12 white burst above the formation for 2 minutes. Aircraft 927 and 899 sustained minor damage.

The weather at the base was 5/10 alto-cumulus with alto-stratus at 8,000 feet with visibility of 10 miles. Temperature at 26,000 feet was -39.

Twenty-six aircraft landed at 1527 hours.

## MISSION NO. 134 - 9 FEB 1945

Three aircraft - 49657, piloted by Lt. Col. Nett; 49024, piloted by 1st Lt. Melvin H. Mooring and 52064, piloted by 1st Lt. Carl O. Bostrom, took off to bomb the MOOSBIERBAUM OIL RIFINERY in Austria by PFF. The three aircraft assembled over the airdrome at 6000 feet. There was no fighter escort. Aircraft 49024 returned early as its PFF equipment malfunctioned.

Two aircraft were over the target at 1224 hours and dropped 4 tons of 500 lb. RDX bombs from 25,500 feet by PFF. No photos were obtained. No enemy aircraft were seen and no flak was encountered over the target. Flak was observed at Bruck, Austria.

Weather: There were 8-9/10s thin cirrus clouds over the base.

Two aircraft landed at 1535 hours.

## MISSION NO. 135 - 13 FEB 1945

At 0904 hours, 21 B-24's, scheduled, took off to bomb the VIENNA SOUTH GOODS YARDS in Austria. The aircraft formed the 1st attack unit and was led by Lt. Col. Douglas M. Cairns, Gp Ops Officer. The Group assembled into Wing formation and was escorted by 40 P-51's from 1200 to 1400 hours. Aircraft 656 could not keep up with the formation because of supercharger trouble and returned to base.

Twenty (20) aircraft were over the primary target at 1235 hours and 19 aircraft dropped 38 tons of 500 lb. RDX bombs visually. Photos indicated that the bombs fell over and to the east of the target, falling in two industrial area destroying several buildings and starting two large fires. No enemy aircraft were encountered. Four Me 109's were seen over Vienna, Austria and four other Me 109's were observed near. Zagreb, Yugoslavia. None of these aircraft attacked the formation.

Weather: Over the base there were a few stratus and cirrus clouds with a visibility of 15+ miles. Twenty-one (21) aircraft landed at base between 1452 and 1515 hours.

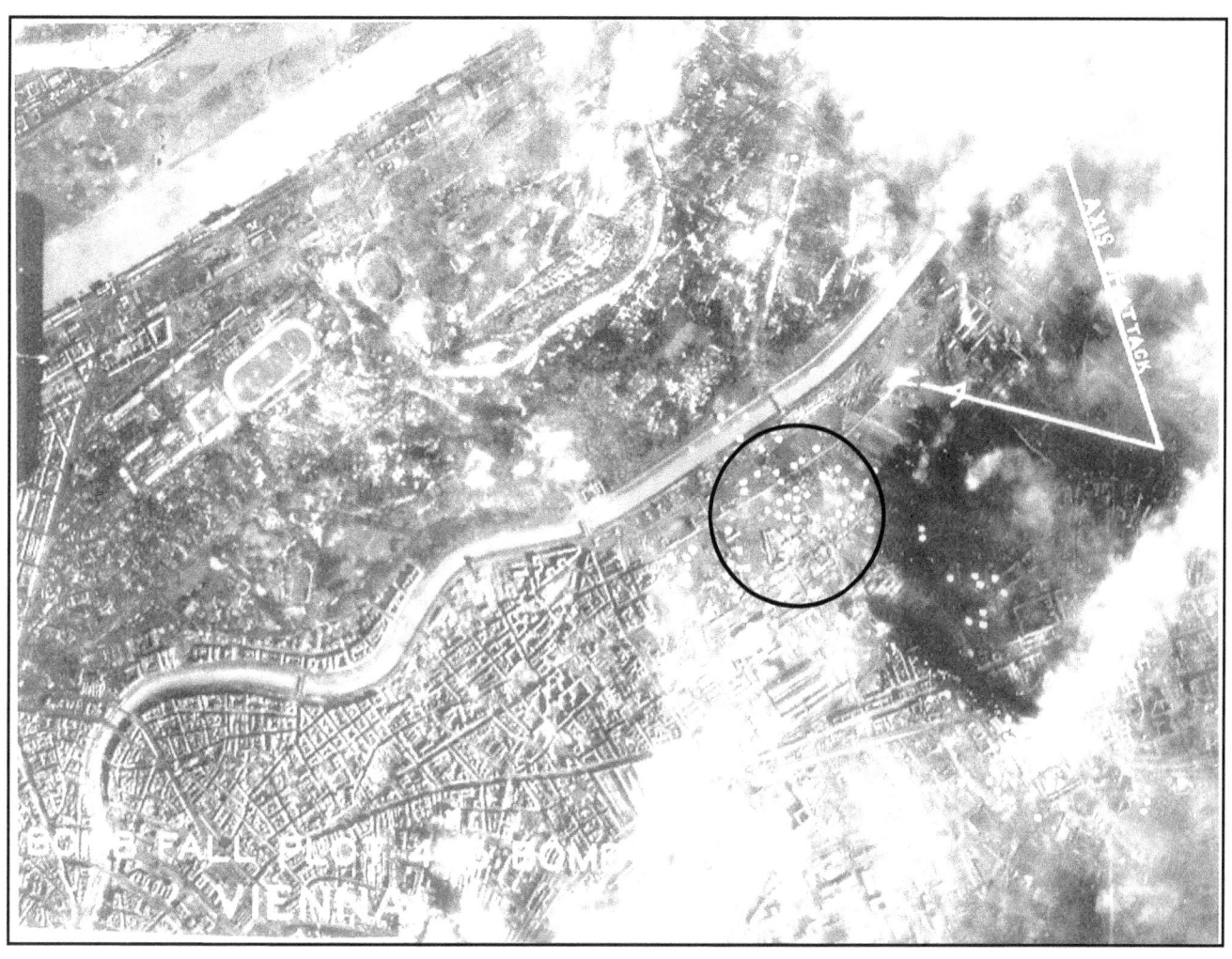

**MISSION 135, 13 FEBRUARY, 1945. TARGET: VIENNA SOUTH GOODS YARDS, AUSTRIA. BOMBS: 500 LB RDX, ALTITUDE 25,000 FEET.**

Austria was the most bombed country in Europe and Vienna was the most bombed city in Europe. There were thirteen targets in Vienna, six oil refineries, four marshalling yards, an ordnance factory, a motor vehicle area, and an industrial area. For this reason Vienna had the most flak cannons. The Danube River can be seen at the very top of the photo, in the left corner. The oil refineries are across the Danube and aren't visible in the photo. The industrial section is in the upper section, below the Danube and in the right half of the photo. The business section can be seen on the lower left. White dots (as shown in and around the circle on the photo) were plotted to show where the bombs fell. This helped to analyze the success of the mission. The "Axis of Attack", as shown, is the direction of the bomb run-northeast.

### MISSION NO. 136 - 13 FEB 1945

At 1131 hours, 21 B-241s, scheduled, took off to bomb the GRAZ MARSHALLING YARD (PT) in Austria. The aircraft formed into one attack unit, which was led by Col. Tomhave, Gp CO. The Group assembled into Wing formation. There was no fighter escort seen and the lead ship contacted the fighter escort by radio 15 minutes before the target and learned that they would be available if needed. Aircraft 779 and 931 returned early.

Nineteen (19) aircraft were over the target at 1459 hours at 22,000 feet and 18 aircraft dropped 36 tons of 500 lb. RDX bombs. The results were good with 20 strikes in the marshalling yard, two strikes on the choke point connecting the two yards and several strikes in the marshalling yard south of the MPI. No enemy aircraft were seen. SAH flak was encountered over the target for 3 minutes and was of the aimed type. Five (5) aircraft received minor flak damage and there were no casualties. It is believed that the slight opposition encountered from flak was due to the element of surprise achieved and from the jamming of enemy radar equipment by the "Panther" aircraft.

Weather: There were a few cirrus clouds over the base with visibility unlimited. The target area was clear of low clouds and visibility was 20 - 30 miles.

Nineteen (19) aircraft landed at base at 1721 hours.

### MISSION NO. 137 - 14 FEB 1945

At 0840 hours, 20 B-24's comprising RED FORCE took off to bomb the VIENNA LOBAU OIL REFINERY in Austria. The formation was led by Major Calvin W. Fite, 828th Ops Officer. The Group assembled into Wing formation and was escorted by 20 P-51's from 1230 to 1330 hours. There were no early returns.

Enroute to the target in the Zagreb, Yugoslavia area, a large front was encountered. Realizing that the formation could not climb above the overcast, the leader elected to go under it rather than to chance scattering the formation by flying through it. This created a problem of climbing to bombing altitude in the relative short time remaining. Aircraft 536 and 859 lost the formation, in this maneuver and were not able to locate the formation before the target was reached, and joined the formation as it rallied from the target after jettisoning their bombs.

Eighteen (18) aircraft were over the target at 1250 hours and dropped 34.5 tons of 500 lb. RDX bombs from 25,500 feet. An effective smoke screen partially obscured the target. Bombing was by PDI. Photos revealed the bombs away and that the target was completely smoke covered. No enemy aircraft were seen. M to IAH flak was encountered over the target for 5 minutes. Several crews reported bursts behind the lead group at regular intervals, indicating that the radar equipment was tracking the chaff streams. Eleven (11) aircraft received minor flak damage.

At the Italian spur, on return, aircraft 51335 contacted the leader reporting that their fuel was running low, two engines being out at that time. The leader gave a bearing to the nearest friendly field and advised the crew to bail out if there was any doubt as to the plane being able to reach the field. Later, a report was received that an aircraft crashed near Amendola airdrome. One crewmember received a broken arm.

Nineteen (19) aircraft landed at base at 1646 hours.

At 0946 hours, 21 B-24's, comprising BLUE FORCE, took off to bomb the same target in Vienna, Austria. Lt. Col. Reeve, 829th CO led the formation. The Group assembled into Wing formation and 40 P-51's escorted the formation from 1250 to 1430 hours. Aircraft 067 returned early.

Shortly after turning on the IP the lead aircraft accidentally released its bombs and 12 aircraft dropped 21.75 tons of 500 lb. RDX bombs from 25,500 feet at 1341 hours, mistaking this for the signal to release their bombs. Photos revealed that the bombs fell in an open field a few miles from the IP, causing no military damage.

Five (5) Me 109's were seen at Vienna, Austria, flying very low and gaining altitude, as if taking off - probably from Zwolfaxing Airdrome. No enemy aircraft were encountered. SIH flak was encountered in the target area and one aircraft received minor flak damage.

Weather: At the base at take off time it was clear. On return there were 10/10s stratocumulus clouds at 1500 feet with tops at 4000 feet and visibility was 7 to 10 miles. Vapor trails persisted at 20,000 feet over Zegrab. There was a 2-4/10s undercast with tops at 15,000 feet over the target area.

Twenty (20) aircraft landed at 1644 hours.

## MISSION NO. 138 - 15 FEB 1945
RED FORCE

At 0903 hours, 18 B-24's took off to bomb the WIENER NEUSTADT MAIN STATION, the primary target. Lt. Col. Atkinson, 831 CO led the formation. Line rendezvous was effected. Thirty P-51's escorted the Group from 1235 to 1419 hours. One aircraft returned early. Aircraft 654, unable to keep up with the formation, lagged behind, turned short of the target and rejoined the formation on the rally.

Sixteen aircraft were over the target at 1302 hours and dropped 32 tons of 500 lb. RDX's from 23,900 feet by PFF. Photos showed the target completely cloud covered. No enemy aircraft were seen; no flak encountered and no damage to aircraft.

There was an overcast at 3,000 feet over the base with 8 mile visibility. Seventeen aircraft landed at 1528 hours. BLUE FORCE.

At 1004 hours, 19 B-24's took off to bomb the WIENER NEUSTADT MAIN STATION, the primary target. Major Harold Pruitt, 830th Co, led the formation. Line rendezvous was effected. Fifty P-51's escorted the Group from 1306 to 1450 hours.

There were no early returns.

Nineteen aircraft were over the target at 1402 hours and dropped 37.5 tons of 500 lb. RDX's by PFF. Photos revealed complete cloud coverage of the target. No enemy aircraft were seen; no flak encountered and no damage to aircraft.

Nineteen aircraft landed at 1630 hours.

## MISSION NO. 139 - 16 FEB 1945

At 0902 hours, 36 B-24's took off to bomb the REGENSBURG/OBERTAUBLING A/C (PT) in Germany. The 1st attack unit was led by Col. Tomhave, Gp CO, the 2nd attack unit was led by Captain Gerald. H. Morris, 831st Asst. Ops Officer and the 3rd attack unit was led by 2nd Lt. Thomas Y. Frase, Jr., 830th pilot. The Group assembled into Wing formation. Several P-38's were sighted at 1135 hours. Approximately 30 P-51's were observed. Escort proceeded the formation over the target, flying low. The escort was last seen at 1515 hours. Aircraft 638 returned early due to no. 3 generator malfunctioning.

Some difficulty was experienced in climbing through the overcast on assembly. Although smoke from previous bombings covered practically the entire airdrome, the target was bombed visually.

Thirty-five (35) aircraft were over the target at 1318 hours and dropped 62.3 tons of 100 lb. clustered frag. bombs from 23,000 feet. Photos revealed hanger target area smoking from bombing by preceding Group and a good bomb pattern of incendiaries starting slightly short (S) of the factory area and continuing into the smoke obscured area to the Airdrome beyond.

No enemy aircraft were encountered. One crew reported, seeing 2 Me 163's over the target flying low and leaving intermittent contrails. On return, 2 FW 190's were seen in northern Yugoslavia. MAH flak was encountered at the target for two minutes. Enroute flak at 46-30N & 13-22E. On return, the formation navigated around this area only to encounter SAH flak a few miles to the west at approximately 46-20E at 1415 hours. The flak bursts appear in groups of fours, with no more than 2-4 gun batteries firing. The formation was at 17,000 feet. The lead aircraft, 657, and 772, flying no. 3 position of the lead box, received direct hits and went down after having collided. After an evasive turn to the right and then one to the left, 772 hit the top of 657, breaking 657 in half right behind the ball turret. Aircraft 657 plunged to earth, crashing near 772, which went down in a slow spiral. A total of 5 chutes were seen to have to have opened, apparently one from aircraft 657 and 4 from aircraft 772. Bombing was accomplished visually.

Weather: At the Base on take off there was an overcast at 4,000 feet with visibility of 15 miles. The overcast continued over the Adriatic Sea where it cleared. The target area was clear.

There were two casualties: Lt. Blair, Co-Pilot of 029 and Lt. Whitney, Navigator of 859. Thirty-two (32) landed at 1644 hours. Aircraft 859 landed at Biferno for fuel.

829th Aircraft 49657

| | |
|---|---|
| P | Col. John P. Tomhave |
| CP | Capt. Richard H. Boehre |
| N | 1st Lt. John L. Carmody |
| N | 2nd Lt. James P. Cahen III |
| B | Major Olen C. Bryant |
| B | 2nd Lt. Marvin A. Woodcock |
| ENG | TSgt. James W. Dixon |
| RO | TSgt Bruce Graves |
| G | SSgt Roy W. Burke |
| G | SSgt Lewis B. Matthews |
| G | SSgt Walter F. Fergus |

829th 47772

| | |
|---|---|
| P | 1st Lt. Carl D. Stockdale |
| CP | 2nd Lt. Arnold M. Mick |
| N | 2nd Lt. William T. Miller |
| ENG | SSgt. Milton Wolfson |
| RO | SSgt. Theodore W. Molek |
| G | SSgt Frank J. Grippo |
| G | SSgt Jesse L. Hall |
| G | SSgt John J. Flynn |
| G | SSgt Early V. Beatty |

Aircraft 772 hit by flak the same time as aircraft 657. Both aircraft went down. One chute seen. Aircraft 657 and 772 apparently hit by flak or collided. Tail section of aircraft 657 came off and the aircraft plunged to earth. One chute seen. Aircraft 772 lost half of wing and spun down slowly. Four chutes were seen. Aircraft 657 received a hit and crashed into aircraft 772. One chute was seen from 657 and three from aircraft 772. Both aircraft were seen to crash on the ground within half mile of each other at 46-31N & 13-42E.

After the raid on Regensburg, February 16,1945, another drama took place through the Brenner Pass. The following is an account as told to Jo Haden Galbraith, daughter of Lt. Robert (Bob) 0. Haden, Navigator from the 831st squadron who passed away in 1995. He was on Glenn Hess's crew and they were on the raid to Regensburg. The target was the Obertraubling Messerschmitt assembly plant. It was the largest plant of its kind in Europe and turned out 200-300 ME-109 fighter aircraft each month.

## PERFECT PASS

### Jo Haden

The crew was on a mission to bomb the Messerschmitt Plant at Regensburg, Germany. As they began the bomb run through heavy flak the number one engine took a direct hit, blowing the prop into the sky and causing the plane to buck like a wild bronco. It was immediately thrown into a severe left bank as the pilot, Lt. Hess, struggled to regain control. Unnerved, and now flying with only three engines, they courageously pressed on toward the target. The bomb bay doors were opened, and within seconds the number two engine was hit, blowing off the turbo charger. Fortunately it did not explode, but the impact caused the plane to bank hard to the right out of control. To make matters worse it threw them into the prop-wash of another bomber, causing the plane to flip upside down. The order was given to bail out, but the centrifugal force caused by the fierce spinning kept the men pinned to the airplane floor and walls frantically trying to pull themselves out of the hatches and waist windows. Caught in a death trap and unable to budge, the crew began their final prayers when the plane (aided no doubt by a little Divine intervention) miraculously righted itself enabling the pilot to pullout of the spin and regain control. Now at 10,000 feet and with a limited amount of fuel the crew was forced to make some quick decisions. They had two choices: fly.to Switzerland, which was doable, or take their chances and try to make it to the allied border in northern Italy. If they landed in Switzerland, a neutral country, they knew they would be interned there for the rest of the war. This did not sit well with the men, as there was no telling how long that might be, possibly years. They were also concerned that they might be classified as M.I.A. (missing in action), causing undo stress on their families. Unable to maintain an altitude higher than 10,000 feet with only two engines, the navigator, Lt. Haden, searched for a route to Italy that would cut through the 15,000 ft. Alpine Mountains. He found it in Brenner Pass, a valley which connects Innsbruck, Austria with Bolzano, Italy.

Brenner Pass is technically at the border between Italy and Austria. The crews always considered it to include the entire valley that snakes through the Alps Mountains with Verona at the South end and Innsbruck at the North end. In places the valley is just wide enough for a river, a road and a railroad. It was a main connection between the Axis. The valley is well over 100 miles long and every foot was heavily defended by 558 large antiaircraft gun installations. Under the best conditions in peacetime a journey through the Alps at that altitude would be considered treacherous. For a crippled bomber low on fuel and being shot at from all sides, it was damn near suicide. To further complicate matters much of northern Italy was still occupied by the Germans, which meant even if they made it through the Alpine Pass in one piece, they would still have a considerable flight over enemy territory.

Fuel was a major concern. Before take off the tanks were topped off at 2750 gallons and the planes were loaded to the hilt with bombs. On the way to the target the group tried to gain as much altitude as possible, consequently burning about 3/4 of the fuel by the time they reached their mark. This meant there might be as little as 600 gallons of fuel left after the run. However, if they made it over the Alps it would be downhill the rest of the way.

Haden calculated that if the Gods were with them (and if they didn't hit a mountain or get blown out of the sky) they would have just enough fuel to eke across the Allied border into Rimini, a coastal town on the Adriatic with an army base and runway. With no time to ponder the idea a vote was taken, and trusting their navigator, the captain and crew opted to take their chances and go for it.

Needless to say it was a harrowing flight through the snow-covered Alps,(pilot Glenn Hess likened it to guiding an elephant through the eye of a needle under fire) but somehow against all odds, their badly crippled plane managed to make it through the Pass, cross the allied border on fumes, and hobble to a stop at the tail end of the Rimini runway. Hess checked the fuel gauge - it was empty.

Stunned and badly shaken by their ordeal, the men crawled out to inspect the plane. Hess recalled: "The plane was so badly shot-up that you couldn't lay your hand anywhere oh it without touching a flak hole. We hadn't been out of the plane more than two minutes when this General came flying down the runway raising all kinds of Hell about us landing on his airstrip. It was a fighter strip and the General was screaming at me to get my f ------ plane off his runway!' I stood there and took his insults for a while until finally exasperated I stopped him by saying 'Sir, would you like to  inspect my plane?' We looked at each other for a moment and then I just walked away. Once he got a good look at it we heard him yell, 'Hell, this thing ain't worth movin!' He then ordered a bulldozer to shove it over the nearest embankment, and that's where it stayed. "

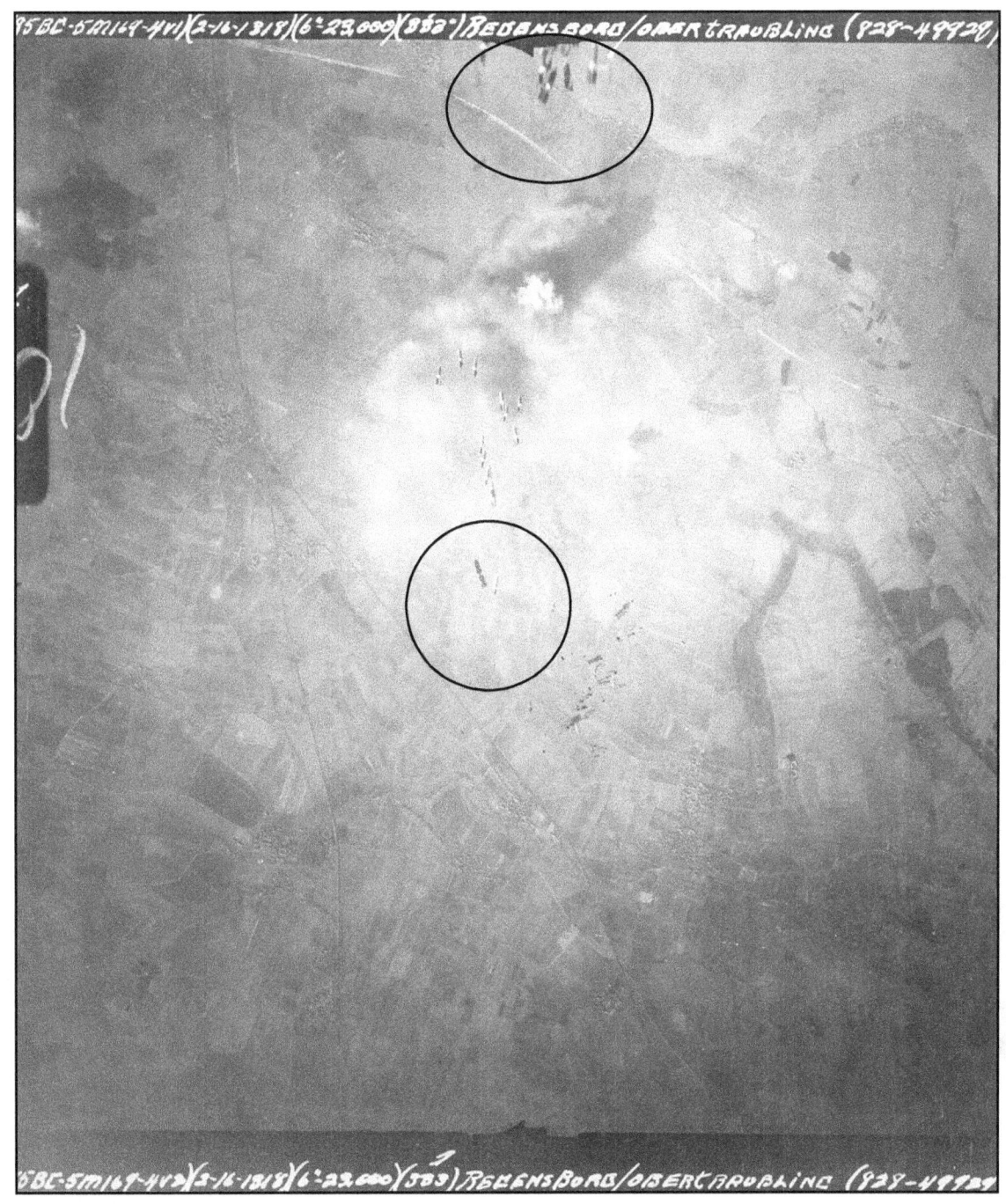

**MISSION 139, 16 FEBRUARY 1945. TARGET: REGENSBURG OBERTRAUBLING ASSEMBLY PLANT (ME-109 FIGHTER PLANES). BOMBS: FRAGMENTATION**

Frag bombs had thin skins and exploded into hundreds of steel splinters when they hit. They were bundled into clusters so they would be heavy enough to drop and clear the plane. Then a small charge would break the bundle apart on the way down. We weren't too happy to drop frags because we were never sure when the bundles would break apart. If a bundle broke apart early, there was a chance of hitting our own plane. In the center of the photo is a bundle still intact, but the bundle at the top of the photo has already broken apart.

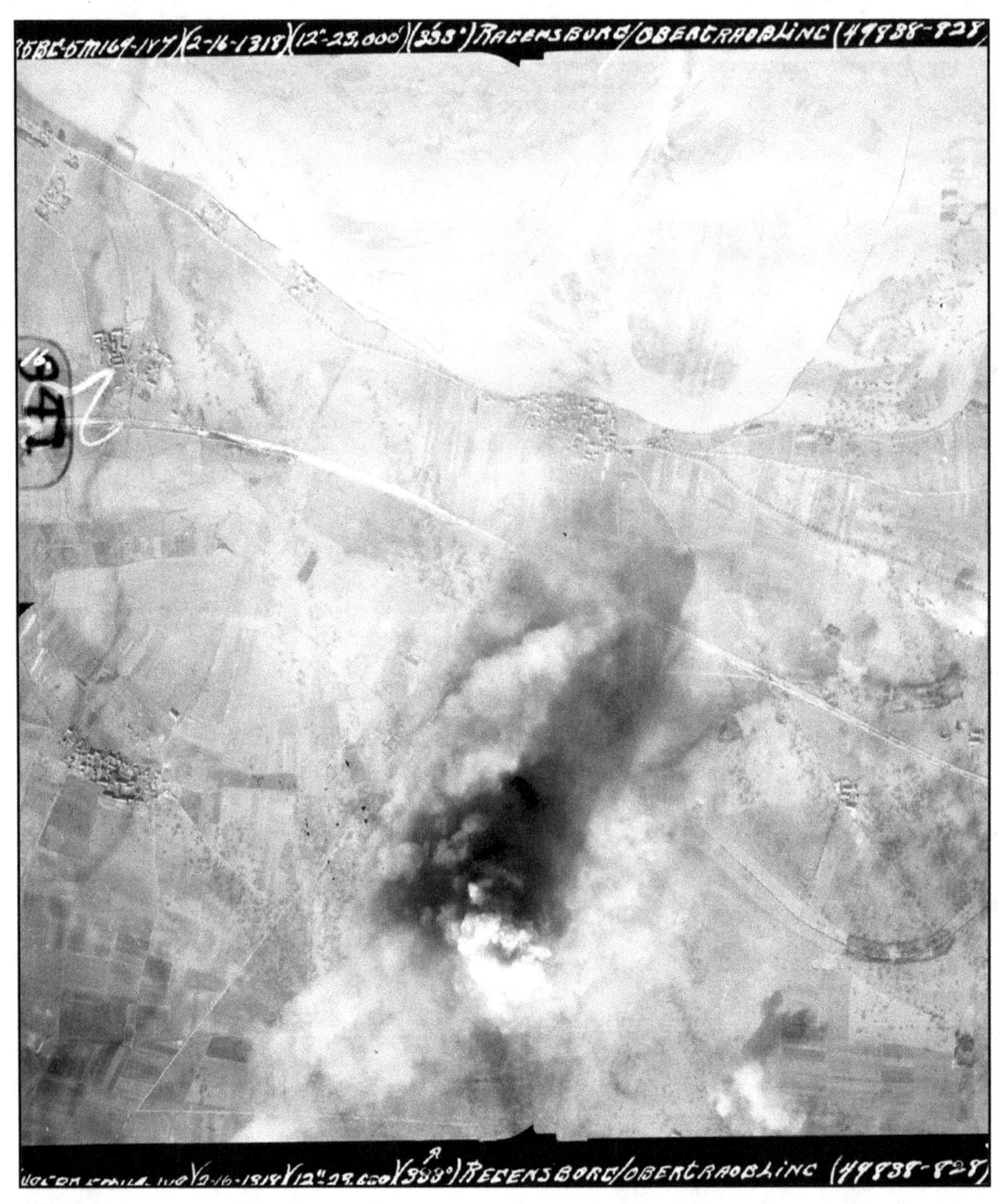

**MISSION 139, 16 FEBRUARY 1945. TARGET: REGENSBURG OBERTRAUBLING ASSEMBLY PLANT (ME-109 FIGHTER PLANES). BOMBS: FRAGMENTATION BOMBS**

Regensburg was a priority target because complete assembly of the Me-109 fighter plane took place at the Obertraubling plant and everything, except for the engine, was manufactured there. It was the largest plant of its kind in Europe, turning out 25-30% of the total Me-109 production.

BRENNER PASS, 30 MARCH 1945. 23,000 FEET. RECONNAISSANCE PHOTOGRAPH OF MARSHALING YARDS, TRENTO, ITALY.

What we called the Brenner Pass was actually the Brenner Valley. It was protected from end to end with flak guns. It was a natural flak alley. The German guns were located mostly on top of the mountains surrounding the valley. It was a shooting gallery for planes flying near there.

## MISSION NO. 140 - 17 FEB 1945

Between 1100 hours and 1219 hours, 27 B-24's (28 scheduled) took off to bomb the POLA HARBOR TARGETS in northern Italy. The 1st attack unit was led by Lt. Col. Douglas E. Cairns, Gp Ops Officer and Capt. James K. Moore, 828th Flt Cmdr led the 2nd attack unit. Aircraft 727 became stuck in the mud on the taxiway and did not take off. The Group assembled into Wing formation. There was no fighter escort.

Twenty-six (26) aircraft were over the target at 1437 hours and dropped 51.25 tons of 500 lb. RDX bombs from 22,500 feet, visually. Results were generally good. A concentration of strikes were obtained on the island containing the dry-docks, shops and administrative buildings. Several strikes were obtained on the infantry barracks, store buildings and ammo depot. No enemy aircraft were seen. MAH flak was encountered over the target for three minutes. Fourteen (14) aircraft received minor flak damage.

Weather: There was an overcast at 1400 feet with tops at 5000 feet. Visibility was 5 miles. There were 3/10 stratocumulus clouds over the target area.
Twenty-five (25) aircraft landed at base at 1608 to 1640 hours. Aircraft 592 landed at Bari with wounded aboard.

## MISSION NO. 141 - 19 FEB 1945

At 0755 hours, 28 B-24's took off to bomb a primary Target in Austria. Lt. Col. Andrus, Deputy Gp Co led the first attack unit and Captain Ward Ritche, 829th Flt Cmdr led the 2nd attack unit. Difficulties were experienced on assembly due to weather conditions over the base. However the Group formation was formed at 0850 hours. Aircraft 536, 694, 438 and 819 returned early. Forty (40) P-38's escorted the formation from 1014 to 1210 hours. Because of difficulties in maintaining formation due to the weather, the leader decided to bomb the 2nd alternate target. Again the formation difficulties prevented bombing and the leader advised the formation that he was heading for POLA, ITALY.

For some reason, several of the aircraft dropped early. Nine (9) aircraft dropped 18 tons of 500 lb. RDX bombs from 23,000 feet at 1320 hours. "D" box made an individual bomb run on Maribor. Five (5) aircraft dropped 10 tons or bombs from 23,000 feet at 1240 hours. Both target were bomb visually. Results were generally poor. Maribor bombs fell in an open area outside the city. No military damage noted. Pola - most of the bombs fell in the bay with one direct hit on a coal quarry. No enemy aircraft were encountered. Near Klagenfurt area, four unidentified aircraft were observed and were believed to be jet propelled. At Jedenburg, 12 single-engine aircraft were observed on a field. Flak at Graz was SIH, at Maribor, it was SIH and at Pola it was SAH flak encountered.

Aircraft 930 crashed landed at Sterporone. Aircraft 410 and 064 landed at Falconara: aircraft 507 landed at Vis; aircraft 067 landed at Lesi; aircraft 834 landed at Ramitelli and aircraft 51812 is missing.

Weather: At the base there were 6-8/10s low clouds at 2000 to 3000 feet. Over the target area were 3-6/10s low clouds at 8,000 feet. At the base on return there were 8-10/10 low clouds with light snow showers.

Seventeen (17) aircraft landed at 1438 hours.

828th Bomb Sqdn Aircraft 51812

| | |
|---|---|
| P | 1st Lt. Thomas J. McKeon |
| CP | F/O Harry B. Christenson |
| N | 2nd Lt. Jerome O. Feldman |
| B | F/O John P. Mulvihill, Jr. |
| RO | Cpl. Roy Wilkerson |
| ENG | SSgt Charles C. Tusk |
| G | Cpl. Elroy C. Meyer |
| G | Cpl. Charles A. Spettel |
| G | Cpl. Thomas L. Barr |
| G | Cpl. Albert S. Teeve |

Formation was scattered badly. No one observed aircraft 812 to leave the formation or no report was heard over the radio. Formation was in the Maribor area at 1240 hours and over Pola at 1320 hours.

## MISSION NO. 142 - 20 FEB 1945

At 0831 hours, 21 B-241s, scheduled, took off to bomb a primary target in northern Italy. The formation was led by Lt. Col. Reeve, 829th CO. Due to unfavorable weather over the primary target, the 1st alternate target, TRIESTE HARBOR INSTALLATIONS were attacked. Aircraft 059, late to take off, was unable to locate the formation and returned at 0950 hours. Aircraft 458, returned before group assembly. The Group formed into Wing formation. There was no definite rendezvous with fighter escort. However P-38's were seen enroute and over the target area. Aircraft 276 returned prior to bombing.

Eighteen (18) aircraft were over the target at 1206 hours and dropped 36 tons of 500 lb. RDX bombs from 22,500 feet. Results showed heavy concentrations of bomb strikes on the harbor installations in the basin causing fires and explosions. The bomb pattern was good. No enemy aircraft were seen. SIH flak was encountered over the target for 3 minutes. One aircraft received minor flak damage.

Eighteen (18) aircraft landed at 1409 hours.

## MISSION NO. 143 - 21 FEB 1945

At 0925 hours, 26 B-24's took of to bomb the VIENNA MATZIEINSDORF MARSHALLING YARD (PT) in Austria. The 1st attack unit was led by major Calvin Fite, 828th CO and the 2nd attack unit was led by Captain Robert L. Brown, 831st Flt Cmdr. The Group assembled at 1025 hours above an overcast at 10,000 feet and were late forming into Wing formation. Thirty (30) P-51's escorted the Group from 1235 to 1500 hours. Aircraft 727 and 067 returned early.

Twenty-four (24) aircraft 48 tons of 500 lb. RDX bombs from 25,000 feet. The bomb run was made visually and aided by PFF. Results: The target was almost completely covered with clouds and smoke. Enough check points were visible to definitely establish that the formation flew over the target. A few strikes were noted several thousand feet west of the marshalling yard.

No enemy aircraft were encountered. At Judenburg, one unidentified single engine fighter was observed flying wide and below the formation. Near the IP, 8 unidentified single-engine aircraft were observed flying below the formation. Several P-51's of the escort were observed to drop their belly tanks and to dive steeply as if to attack. IAH flak was experienced over Vienna for 5 minutes. Nine (9) aircraft received minor flak damage.

Weather: At the base at take off there was scattered overcast with breaks at 3,000 feet with light snow showers. Is was clear over the target area and visibility was 20 miles.

Twenty-four (24) aircraft landed at 1636 hours.

## MISSION NO. 144 - 22 FEB 1945

At 0816 hours, 30 B-24's (31 scheduled) took off to a target in Germany, according to AF plan CLARION. Col. John B. Cornett, Gp CO led the formation. Aircraft 059, taking off at 0928 hours, was unable to join the Group, formation and flew with the 464th Gp. Group by-passed wing rendezvous. Fifty P-51's provided fighter escort from 1150 hours until the group returned to the Italian coast. Aircraft 638, 829 and 728 returned early.

The mission was aborted due to weather. Over the southern slopes of tile Alps, clouds based on the peaks were encountered. The leader elected to climb the formation through these clouds rather than to chance crossing the Alps at low altitude through doubtful weather. The formation broke through to the told of the cloud layer at 20,000 feet. Prior to the climb the formation had been very good but it became somewhat scattered during the ascent. After reassembling the formation continued on course, observing that the density of the undercast became greater to the north in the vicinity of the target. After being informed by the Wing Leader, that he was abandoning the primary target, the Group leader decided to return to the. Udine area and bomb the alternate target there, which was observed clear at 1135 hours. The formation turned at 1223 hours after having reached an altitude of 21,500 feet. The undercast precluding the bombing of any targets in Southern Germany. On return, it was found that the undercast had moved to the south, preventing bombing of targets in northern Italy. Aircraft 059 bombed Casarsa with the 464th, dropping 2 tons of 500lb RDX's at 1255 hours from 15,000 feet.

No flak was encountered. One Me 163 was observed in the Udine area at 1140 hours flying alone at some distance from the formation. Eight unidentified aircraft, probably Me 262's were seen at 1330 hours at 47-OON and 13-OOE. These were described as having engine nacelles and single tails and were flying wide and high in 2 formations of 3 and 5 aircraft.

Over the base there was 7/10 stratocumulus at 3,000 feet. Twenty-seven aircraft landed at 1504 hours.

## MISSION NO. 145 - 23 FEB 1945

At 0921 hours, 28 B-24's took off to bomb a primary target in Austria. Prevented from bombing the PT because of 10/10s undercast, the 1st alternate target at BRUCK, Austria was attacked. The 1st attack unit was led by Major Pruitt, 830th CO and the 2nd attack unit was led by Major Cummings, 829th CO. Wave rendezvous was effected. Twenty P-51's escorted the formation from 1215 to 1400 hours. Aircraft 296, 410 and 414 returned early. After the lead aircraft turned back, Captain Jacob Disston, 830th Ops Officer took lead of the formation. Eighteen (18) were over the target at 1330 hours and dropped 45 tons of 500 lb. RDX bombs from 20,500 feet by PFF and visual means. Results: Missed the target.

No enemy aircraft were encountered. Two Me 109's were observed at 1244 hours apparently attempting to attack a red tail B-24 which had turned back. Five P-51's came in and drove the Me 109's away. MAH flak was encountered over Bruck for 2-3 minutes. Aircraft 838 flying in no. 4 position in "B" box received a direct hit at 1311 hours and burst into flames. Seven (7) men were definitely seen to have left the aircraft. A total of six chutes were seen to have opened, one which was burning. Unless trapped in the burning aircraft, all crew members had ample time to bail out. Aircraft 48764 had its ailerons and hydraulic system shot out. Pilot Adams called Major Cummings, the formation leader, requesting a heading to the nearest friendly territory. Major Cummings led him over Zara, Yugoslavia at 1431 hours. Pilot reported that he was going to bail his crew, rather than chance the crossing the Adriatic with a questionable supply of gas and his aircraft severely damaged. It was reported from AF that the entire crew bailed out over Prkos and that all were safely evacuated to Bari, Italy in a few hours. Fifteen (15) aircraft received flak damage, 2 major damage and 13 minor damage.

Weather: It was clear over the base. Over the target area there were thin altocumulus clouds at 16,000 feet. On return to the base there was an overcast at 3200 feet. Twenty-three (23) aircraft landed at 1600 hours. Pilot Prishkorn landed last as his aircraft had the rudder controls shot away.

828Th Bomb Sqdn Aircraft 49838

| | |
|---|---|
| P | 1st Lt. Robert G. Ware |
| CP | 2nd Lt. William T. Ryan |
| N | F/O Jack Hubbard |
| B | 2nd Lt. Arthur I. Deahn |
| RO | Sgt. Merle E. Shields |
| ENG | SSgt. William L. Sniegowski |
| G | SSgt. Edward D. Hope |
| G | Sgt. James L Lazarakis |
| G | Sgt. Henry J. Ring |
| G | Sgt. Robert J. Yates |

## MISSION NO. 146 - 24 FEB 1945

At 0909 hours, 28 B-24's took off to bomb a primary target in Italy. The mission was non-effective due to weather. Lt. Col. Andrus, Deputy Gp CO, led the formation. Form-up of boxes was accomplished over the field. Some difficulty was experienced on assembly when the 461BG of the 49 BW assembled in the area assigned to the 485BG. However, assembly was completed in good formation at 0946 hours at 9,000 feet. Wing formation was good. Thirty P-51's escorted the Group, departing at 1315 hours. Aircraft 931 and 929 returned early.

The Wing formation was very good until weather was encountered over the Adriatic, enroute. Over the Adriatic, every effort was made to get around the weather. At 1255 hours, The leader decided to abort the mission as the weather precluded the bombing of the primary target and any alternate target in the area.

Weather over the base at take off was clear and visibility was 15 miles. Heavy vapor trails formed over 18,000 feet. Twenty-six aircraft landed at 1433 hours.

## MISSION NO. 147 - 25 FEB 1945

At 0821 hours, 28 B-24's took off to bomb the LINZ BENZOL PLANT in Austria. Lt. Col. Cairns, Gp Ops Officer led the formation. Assembly was hindered by weather conditions - there being a deck of 8-10/10s clouds at 2500 feet. There was no rendezvous with other Groups. Forty (40) to fifty (50) P-51's escorted the formation from 1132 to 1400 hours. Aircraft 458 and 416 returned early. The bombing run was by PFF as the target was smoke covered from smoke generators and previous bombing.

Nineteen (19) aircraft were over the target at 1309 hours and dropped 37.5 tons of 500 lb. RDX bombs from 23,300 feet. A few strikes were within the confines of an ordnance plant. Very little military damage accomplished.

No enemy aircraft were encountered. At 1250 hours one Me 109 was sighted near Lake Chiem and was attacked by P-51's. IAH flak was encountered over the target for five minutes. Aircraft 930 received major flak damage and 13 aircraft received minor flak damage. There were no casualties.

Weather: At the base at take off there were 8-10/10s low clouds at 2500 feet with visibility 9 miles. Over the target there were 2-4/10S thin cirrus clouds at 23,000 feet. At the base on return there were 47/10s clouds coverage.

Twenty-six (26) aircraft landed at 1540 hours.

## MISSION NO. 148 - 27 FEB 1945

At 0921 hours, 28 B-24's took off to bomb the AUGSBURG MARSHALLING YARD in Germany (PT). The 1st attack unit was led by Major Cummings, 829th Ops Officer and the 2nd attack unit was led by 2nd Lt. Richard E. Juday, 830 Flt Cmdr. Wing formation was late. Thirty-five (35) P-51's escorted the formation from 1310 to 1425 hours. There were no early returns. Aircraft 049 turned back at 1213 hours.

Twenty-seven (27) aircraft were over the primary target at 1349 hours and dropped 67 tons of 1000 lb. RDX bombs from 24,700 feet by PFF became of smoke and haze over the target. Results: Approximately 25 strikes near the MPI. Smoke prevented accurate damage assessment.

IAH flak was encountered over the target for 5 minutes. Five aircraft received major flak damaged and 13 aircraft received minor flak damage. There were no casualties and no enemy aircraft were seen.

Weather: It was clear over the base at take off. Over the target there were 5-7/10s patchy low clouds.

Aircraft 52727 landed at Lesi Airdrome for gas and returned after.

Twenty-six (26) aircraft landed at 1650 hours.

## MISSION NO. 149 - 28 FEB 1945
RED FORCE

At 0953 hours, 17 B-24's took off to bomb the ORA MARSHALLING YARD in northern Italy. The formation was led by Major Fite, 828th CO. Wing rendezvous was accomplished. There was no fighter escort. Aircraft 517 failed to take off.

Seventeen (17) aircraft were over the primary target at 1410 hours and dropped 42.5 tons of 500 lb. RDX bombs from 23,300 feet, visually. Results: A few strikes were obtained in the southern extremity of a small marshalling yard. Fifteen strikes were made in the northern half of the target near the MPI and three strikes were made on the northern choke point.

No enemy aircraft were encountered. At 1408 hours, 5 unidentified fighters were observed in the target area. They were probably Macchi 202s. MAH flak was encountered for 1 - 2 minutes over the target.

Weather: There were 4/10s cirrus clouds at 20,000 feet over the base. Over the target there were 3/10s cirrus clouds at 25,000 feet. On return it was clear over the base.

Seventeen (17) aircraft landed at 1636 hours

BLUE FORCE

At 1018 hours, 18 B-24's took off to bomb the ORA MARSHALLING YARD in northern Italy. The formation was led by Lt. Col. Atkinson, 831st CO. Wing rendezvous was effected. There was no fighter escort. Aircraft 394 returned early.

Seventeen (17) aircraft were over the primary target at 1429 hours and dropped 42.5 tons of 500 lb. RDX bombs from 23,000 feet. Results: Ten strikes were obtained in the north end of the yard and the rest of the target was missed. No enemy aircraft were seen. MAH flak was encountered for 2 minutes. Two aircraft received flak damage and eight aircraft received minor flak damage. On aircraft 829, Sgt. C. M. Bean was wounded. Aircraft 024 landed last, having been badly damaged by flak. SSgt. James Kan, Flt Eng., spliced a rudder control cable, which had been severed by flak, while in flight. Seventeen (17) AC landed at 1648 hours.

# UNIT HISTORY, 485TH BG (H) 1 MARCH 1945 TO 31 MARCH 1945

During the month of March, the 485th BG flew 20 combat missions, making a grand total of 169 missions as of 31 March 1945. 1,288 tons of bombs were dropped. The greater part of this tonnage was dropped by visual means.

On these missions, five crews and their aircraft were lost from enemy action. Nothing more has been heard on these crews.

| | |
|---|---|
| Aircraft on hard - 1 March 1945 | 60 + 3 WWR's |
| Aircraft on hand - 31 March 1945 | 61 + 3 WWR's |
| Aircraft lost to operational accidents | 1 |
| Aircraft lost to non-operational accidents | 1 |
| Aircraft lost to combat | 8 |
| Aircraft transferred to 15AF Service Command | 19 |
| Aircraft gained (new) | 15 |
| Aircraft repaired by 15 AF SC | 12 |

On 22 March 1945, Col. John B. Cornett, Gp CO, since 17 February 1945, was reported MIA over Austria. Col. Cornett's aircraft was hit by flak over the primary target in Vienna. When last seen his aircraft was headed east under control and the Col. informed the formation that he would attempt to land behind the Russian lines. To date, no word has been heard from his aircraft.

The Group Operations Officer, Lt. Col. Douglas M. Cairns was appointed Gp CO on 23, March 1945.

AWARDS

| | |
|---|---|
| Legion of Merit | 2 |
| DFC | 21 |
| Bronze Star | 3 |
| Purple Hearts | 9 |

## MISSION NO. 150 - 1 MARCH 1945
RED FORCE

At 1021 hours, 16 B-24's took off to bomb a primary target in Austria. Weather conditions precluded the bombing of the PT and the AF no. 4 alternate target, ST. POLTEN, MARSHALLING YARD in Austria was bombed. Major Pruitt, 830th CO led the formation.

Aircraft 446 slipped into a ditch, became stuck and was unable to take off. Aircraft 644 returned early. There was no Wing rendezvous. Thirty-five P-51's escorted the formation from 1305 to 1545 hours.

Fifteen (15) aircraft were over the target at 1435 hours and dropped 30 tons of 500 lb. RDX bombs from 21,500 feet by PFF. Photos revealed that the entire target area was completely cloud obscured. No enemy aircraft were seen, no flak encountered and there was no damage or casualties.

Fifteen (15) aircraft landed at 1720 hours.

BLUE FORCE

At 1044 hours, 15 B-24's took off to bomb a primary target in Austria. Because of the flux gate compass malfunctioning while on the bomb run, to the PT, this target was abandoned and the 2nd alternate target AMSTETTEN MARSHALLING YARD in Austria was bombed. Col. Cornett, Gp CO led the formation. There was no wing rendezvous. Twenty (20) P-51's escorted the formation from 1314 hours to 1450 hours. There were no early returns.

Fifteen (15) aircraft were over the target at 1448 hours and dropped 30 tons of 500 lb. RDX bombs from 24,100 feet by PFF. Photos showed bombs away with the ground completely obscured. No enemy aircraft were seen. No flak was encountered over the target. Enroute, SAH flak was encountered for approximately a half minute. Since this location is near the Russian battle lines, and since this flak was not shown on the latest flak maps, it is very likely it was railroad flak, or possibly flak guns which may be serving the dual purpose for the Germans as anti-aircraft and anti-tank defense. Two aircraft received minor flak damage.

Weather: Over the base there were 6/10's cirrus clouds and 15+ miles visibility. Over the target there was 10/10s stratocumulus with breaks.

Fifteen (15) aircraft landed at 1735 hours.

## MISSION NO. 151 - 2 MARCH 1945

At 0903 hours, 28 B-24's took off to bomb the LINZ SOUTH MARSHALLING YARD (PT) in Austria. The 1st attack unit was led by Lt. Col. Andrus and the 2nd attack unit was led by Captain Edward E. Richie, 829th Flt Cmdr. A line rendezvous was effected. Aircraft 393 and 336 failed to assemble with the formation. Also craft 908 and 507 returned early

At 1252 hours aircraft 52064, flying no. 2 position in "C" box and aircraft 52664, flying no. 7 position in "B" box collided, resulting in the loss of both aircraft. A total of six chutes were seen to have opened.

Twenty (20) aircraft were over the target, releasing bombs while in a turn at 1345 hours and dropped 40 tons of 500 lb. RDX bombs from 23,400 feet. Photo's revealed broken clouds over the target which almost obscured the entire area. A good bomb pattern fell in a cleared area approximately 29,000 feet east of the PFF aiming point. Aircraft 758, unable to maintain formation after losing no. 3 engine, dropped 2 tons of bombs from 19,6000 feet. It is believed that the bombs fell over the target into the town of Burghausen. No enemy aircraft were seen.

MAH flak was encountered over the PT for three minutes. Four aircraft received flak damage: One major damage and three minor damage. There were no casualties.

Two aircraft are missing. Aircraft 458 turned back near Lake Chiem at 1220 hours with an engine out. Aircraft 059 was heard calling Big Fence on return for a fix for the isle of Vis. The formation at that time was over the Yugoslavian mainland. It is believed that one of these aircraft is at Zara, Yugo, and the other at Vis, though no confirmation has been received.

Weather: At take off it was clear over the base. Over the target there was a 8-910's undercast.

Twenty aircraft landed at 1623 hours.

828th Bomb Sqdn Aircraft 40458
P     2nd Lt. Richard T. Loudon
CP    2nd Lt. John G. Trieber
N     F/O Francis Ryan
B     2nd Lt. Clyde E. Herbold
ENG   Cpl. William C. Washburn
RO    Sgt. William R. Grance
G     Cpl. Francis A. Bain
G     Cpl. Roy L. Wason
G     Cpl. Philip W. Lindler
G     Cpl. Andrew E. Belloin
G     SSgt Fredrick F. LaPlante

Aircraft 458 turned back at Lake Chiem with no. 3 engine feathered at approximately 1220 hours.

829th Bomb Sqdn Aircraft 52064
P     1st Lt. Earl W. Pooley
CP    1st Lt. James Michalaros
N     2nd Lt. George A. Fuccillo
N     2nf Lt. Albert C. Griffin
B     1st Lt. Adam L. Welger
ENG   TSgt. Charles W. Jones
RO    TSgt. Lavern R. Krueger
G     SSgt. John N. Magness
G     SSgt. Walter J. Kuszler
G     SSgt. Walter L. Broker
G     SSgt. Peter D. Lambros

828th Bomb Sqdn Aircraft 52644
P     2nd Lt. Carl W. Langley
CP    2nd Lt. Richard V. Miller
N     F/O William J. Hofemeister
ENG   Cpl. Paul E. Schultz
RO    Cpl. Henry Koprowski
G     Cpl. William S. Kaukas
G     Cpl. George L. Taylor
G     Cpl. Doyle G. Summer
G     Sgt. Leegrand H. Loller

Pilot Hicks – saw two aircraft after they had hit. Six chutes were definitely seen. Pilot Cotton – Aircraft 644 seemed to lose power, dropped back from no. 7 position in "B" box, turned left and rushed into aircraft 064 flying in no. 2 position in "C" box. Two chutes were seen and probably one more. The time was 1252 hours. Pilot Larkin – Aircraft #644 slid back, lost left part of stabilizer, then turned up on right wing and collided with aircraft #064. One aircraft broke-up and went down in a flat spin, the other breaking into pieces.

## MISSION NO. 152 - 4 MARCH 1945

At 0739 hours, 28 B-24's took off to bomb the SZOMBATHELY SOUTH MARSHALLING YARD in Hungary. The formation was led by Lt. Col. Douglas Cairns, Gp Ops Officer. Due to weather conditions over the base, the Group assembled 30 miles west of the field at 0470 hours and at 11,000 feet. Rendezvous with the 460th BG was not effected. There was no actual rendezvous with the fighter escort. Fighters called for the target time and evidently flew directly to the target, where they were first observed in any number. A few P-38's and P-51's were observed enroute. The fighters were last seen at 1322 hours.

Aircraft 801 at 1130 hours turned back and although no planned bomb run was made, the aircraft jettisoned 2 tons of bombs and some hit were observed on a rail line. Aircraft 801 landed at 1410 hours.

With a bomb run of only 45 seconds, 27 aircraft were over the primary target at 1242 hours and 26 aircraft dropped 52 tons of 500 lb. RDX bombs from 23,500 feet, visually. Photos revealed a good bomb pattern on the southern section of the target hitting both MPI's and hits west of the yard and in the center of town.

Weather: On take off over the base there were 9/10 cumulus clouds and strato-clouds at 4,000 feet. Over the target, there were 3/10s low clouds. Same weather was encountered over the base on return.

Twenty-six (26) aircraft returned to base and 25 aircraft landed at 1443 hours. Aircraft 51931 landed at Zara to refuel and returned later. Aircraft 49393 was observed to lose altitude over the target and take up a heading to the east, escorted by several P-38's. This aircraft, later, reported by radio that it had landed behind the Russian lines at Szedebed.

## MISSION NO. 153 - 8 MARCH 1945
RED FORCE

Eighteen (18) B-24's took off to bomb the PORTO NUOVO MARSHALLING YARD at Verona, Italy (PT). The formation was led my Major Fite, 828th CO. Wing formation was not affected and there was no fighter escort. Aircraft 779 returned early.

Seventeen (17) aircraft were over the primary target at 1500 hours and dropped 33 tons of 500 lb. GP bombs from 24,200 feet visually. Results: Ten strikes were in the yard near the MPI and direct hits were noted on the Railroad workshops.

Weather: Over the base there were 7/10s cumulus clouds at 2500 feet. Over the target there was 1/10 strato-cumulus.

Seventeen (17) aircraft landed at 1651 hours.
BLUE FORCE

Eighteen (18) aircraft took off to bomb the same primary target. Major Cummings, 829th Ops Officer led the formation. On take off, the life raft on his aircraft released and hung on the tail assembly fouling the controls. The aircraft managed to stay airborne and gained altitude slowly after the bombs were slavoed near the field. It left the traffic pattern and circled to the right apparently under control, but with the life raft still attached. At 1210 hours, this aircraft crashed approximately 10 miles northeast of Nature field. Four crewmen bailed out safely and seven were killed in the crash. Captain Harold F. Cline, asst. Ops Officer of the 829th took over the lead of the formation. No Wing rendezvous was effected and there was no fighter escort. Aircraft 656, 410 and 438 returned early.

As the formation approached the northern Italian coast, an undercast began to build up, becoming solid over the Alps. As the IP was approached, the leader decided to return to the base, believing that the weather conditions would prevent the bombing of any targets in northern Italy. The formation turned back at 1516 hours. On return, the weather over the north Adriatic prevented the bombing of any of the AF coastal alternate targets, in northern Italy.

Fourteen (14) aircraft returned 28 tons of bombs, landing at 1710 hours. The weather was the same as for Red Force, except for the weather over the target, which had 8-9/10s strato-cumulus cloud coverage.

## MISSION NO. 154 - 9 MARCH 1945
RED FORCE

At 0936 hours, 18 B-24's took off to bomb a primary target in Yugoslavia. Encountering weather in that area, which precluded bombing the primary target, the formation, attacked the GRAZ STN/Y in Austria. The formation was led by Lt. Col. Atkinson. The Group did not form into Wing formation. Thirty to forty P-38's escorted the Group from 1142 to 1400 hours. There were no early returns. At Toplice, the leader decided to attack Graz.

Eighteen aircraft were over the target at 1332 hours and dropped 36 tons of 500 lb. GP's from 23,000 feet by PFF and visual. Photos revealed that the target was almost completely cloud-covered, and bomb strikes obscured by the clouds.

At take off time, the weather over the base was 2/10 coverage - low at 2,000 feet and tops at 35,000 feet, and visibility - 10 miles. The target area had 7-9/10 undercast. No enemy aircraft were seen. MAH flak was encountered over Graz for 2 minutes. Aircraft 409 received major flak damage. There was one casualty - Cpl. M.L. Holcomb, tail gunner on aircraft 409.

Eighteen aircraft landed at 1528 and 1553 hours.

### BLUE FORCE
Seventeen B-24's took off to bomb the same target as Red Force. Blue Force was led by Major Pruitt, 830th CO. Aircraft 758 returned immediately after take off syphoning gas. There was no Wing rendezvous and no fighter escort. In the vicinity of Toplice, it became evident that the primary target could not be bombed visually. The leader decided to try for the 1st alternate target, which was briefed for visual or PFF bombing. Unable to locate that target on the scope, the formation made an attempt to attack the 3rd alternate target by visual means. This target was also cloud covered. Next, it was decided to bomb targets of opportunity. All targets of opportunity were obscured by clouds.

Sixteen aircraft landed with bombs at 1624 hours.

## MISSION NO. 155 - 12 MARCH 1945
At 0924 hours, 39 B-24's took off to bomb the FLORISDORF OIL REFINERY in Vienna Austria. Col. Cornett led the RED FORCE of 19 aircraft and Lt. Col. Andrus led the BLUE FORCE of 20 aircraft. Line rendezvous was accomplished. Forty - P-51's escorted the Group from 1240 to 1440 hours. Aircraft 863 returned early.

Nineteen aircraft of BLUE FORCE were over the primary target at 1346 hours and 18 aircraft dropped 35.5 tons of 500 lb. RDX's from 23,050 feet.

Nineteen aircraft of BLUE FORCE were over the primary target at 1347 hours and 18 aircraft dropped 36 tons of RDX's from 24,000 feet.

Bombing was by PFF. Photos showed only bombs away and the target area completely cloud covered.

At take off there were a few low and middle clouds with visibility of 8-10 miles. In the target area there was 10/10 cloud coverage with tops to 12-14,000 feet. Temperature at 23,000 feet was -28 degrees C.

No enemy aircraft were seen. M-IAH flak encountered over the target for minutes. Flak was described as being a combination of barrage and tracking type. BLUE FORCE reported most of the flak burst being in the chaff stream of RED FORCE. Six aircraft received minor flak damage. There were no casualties. Jamming of Channel B (VHF) was experienced over enemy territory.

Thirty-seven aircraft landed at 1606 and 163 hours. Aircraft 656 landed at Zara, Yugoslavia after experiencing trouble with no. 3 and no. 4 engines. Crew left aircraft and returned with 47th Wing aircraft to base.

## MISSION NO. 156 - 13 MARCH 1945

Twenty-eight (28) B-24's took off to bomb the REGENSBURG SOUTH-EAST MARSHALLING YARD (PT) in Germany. The formation was led by Major Fite, 828th CO. The group assembled into Wing formation. Forty (40) P-38's escorted the formation from 1235 to 1450 hours. Aircraft 931 returned early.

Twenty-seven (27) aircraft were over the target at 1309 hours and dropped 54 tons of 500 lb. GP bombs from 22,000 feet. Photos revealed that at the bombs away, a layer of clouds completely obscured the ground. No enemy aircraft were seen and no flak was encountered. PANTHER operators recorded only one radar interception at 1255 hours.

Weather: At take off, over the base, there was 1/10 altostratus at 8,000 feet. The target area was covered by 9/10s alto stratus clouds.

Twenty-seven (27) aircraft landed at base between 1557 and 1630 hours.

## MISSION NO. 157 - 14 MARCH 1945

At 0914 hours, 38 B-24's took off to bomb the NOVE SAMKY MARSHALLING YARD (PT) in Hungary. Lt. Col. Atkinson led the RED FORCE of 19 aircraft and Major Dolim, Deputy CO of the 829th led the BLUE FORCE of 19 aircraft. The two force assembled into the bomber stream in line at 13 minute intervals. There was no fighter escort. Seven P-38's were seen at 1235 hours and 10 minutes later 6 P-51's were seen. Aircraft 694, 407, 474, 394 and 528 returned early.

Seventeen (17) aircraft of BLUE FORCE were over the target at 1351 hours and dropped 42 tons of 500 lb. RDX bombs from 21,300 feet, visually. RED FORCE made a second bomb run over the target. Sixteen (16) aircraft were over the target at 1413 hours and 11 aircraft dropped 27.5 tons of bombs from 21,600 visually. "C" box did not drop their bombs, being unable to identify the target. Results: A good bomb pattern in the north-west end of the yard by BLUE FORCE. The bombs of RED FORCE fell in the south-east end of the yard. No enemy aircraft were seen. S-MAH flak was encountered at the target and 6 aircraft received minor flak damage.

Weather: Over the base at take off, it was clear, and on return there were 2/10s cumulus. Over the target there were a few cirrus clouds.

Twenty-six (26) aircraft landed at base at 1646 hours. The 27th aircraft to land crashed on the runway, blocking it, which caused 5 aircraft to land at Pantanella. These aircraft returned later to base. Aircraft 515 landed at Zara. Also aircraft 517 and 899 landed at Zara, refueled and returned later to base.

## MISSION NO. 158 - 15 MARCH 1945

At 0758 hours, 20 B-24's, comprising RED FORCE and led by Col. Cornett, Gp CO, took off to bomb a primary target in Austria. Thirteen aircraft, comprising part of BLUE FORCE, took off to bomb the same target. Seven other aircraft of this force were delayed in taking off due to aircraft becoming stuck and blocking the taxiway in the 831st area. The last of these aircraft were airborne at 0953 hours and were led by Major Pruitt, 830 CO. Line formation was effected by both forces. Sixteen P-38's escorted from 1152 to 1245 hours. Aircraft 059, 038, 638 and 727 returned early.

Before take off, Wing HQ advised that all targets would be bombed by visual means. RED FORCE did not drop their bombs on the target because the primary target was covered by smoke from smoke generators and 4/10 cloud coverage. RED FORCE made a 180 turn to the right and headed for the 1st alternate target, which was 10/10 cloud, covered, preventing visual bombing. RED FORCE made another 180 turn to the left, returned to the Vienna area and then to home base. BLUE FORCE abandoned the primary target and followed RED FORCE over St. Polten and identified the marshalling yard, but had insufficient time for bombing. Circling to the north and returning, BLUE FORCE made an visual attack on the yard at 1243 hours. Eighteen aircraft dropped 43 tons of 500 lb. RDX's from 23,500 feet. Results were excellent with a pattern of bombs on the western end of the yard.

No enemy aircraft were seen and no flak was encountered by either force.

On take off there were 5/10 cirrus clouds and 1/10 altocumulus clouds over the base. There were 2 to 4/10 cumulus clouds in scattered patches but clear in the immediate vicinity of the target.

Four aircraft from RED FORCE landed at Zara, Yugoslavia for gas: 410, 067, 049 and 474; all except 474 returned later to base.

Thirty-two aircraft landed at 1530 hours.

## MISSION NO. 159 - 16 MARCH 1945
RED FORCE

At 0743 hours, 19 B-24's took off to bomb a primary target in Austria. Because of weather conditions which prevented the bombing of the PT visually, the AMSTETTEN MARSHALLING YARD in Austria was bombed. Lt. Col. Cairns, Gp Ops Officer, led the formation. There was no rendezvous with Wing and there was no fighter escort. Aircraft 532, 367, 863, 724 and 059 returned early. Aircraft 528 joined the 465th BG until the turn-point of Fuime was reached. Then the aircraft joined the 376 BG and dropped 8 bombs at 1213 hours from 23,500 feet near Fanhagen.

Thirteen (13) aircraft were over the target at 1200 hours and dropped 25.75 tons of 500 lb. RDX bombs from 23,900 feet, visually. Photos revealed a poor bomb pattern extending 2,000 feet north to the yards with only one hit in the yards.

BLUE FORCE

At 0806 hours, 10 B-24's took off to bomb the sane target assigned to RED FORCE The formation was led by Lt. Col. Andrus, Deputy Gp. CO. The force joined the 460th BG. There was no fighter escort. Aircraft 507, 504, 276, 438 and 834 returned early.

An attempt was made to bomb the marshalling yard but the bomb run was too short. The formation did a 360 turn and made a second bomb run. Fourteen (14) were over the target at 1214 hours and thirteen (13) aircraft dropped 36 tons of 500 lb. RDX bombs, visually. Photo's showed that some bombs fell in the yard and some fell short of the yard.

Weather: At the base, on take off, there was a 6/10s altocumulus over cast. Over the target, it was clear. On return to the base, there were 8/10s cumulus with scattered fractostratus and light showers.

Neither force encountered or saw any enemy aircraft, and no flak over the target. RED FORCE experienced SAH flak, about 20 bursts, in an uncharted flak area. One aircraft received minor damage and there were no casualties.

On return, the formation was hindered somewhat by local weather conditions, and had to circle south of the spur area for several minutes before the thunder showered at the base cleared.

Twenty-six (26) aircraft landed at 1447 and 1525 hours. Aircraft 441 landed at Zara, refueled, and returned to base.

## MISSION NO. 160 - 19 MARCH 1945

At 0813 hours, 19 B-24's (BLUE FORCE) took off to bomb the MUHLDORF MARSHALLING YARD in Germany. The formation was led by William Ceely, 831st Ops Officer. At 0829 hours, 21 B-24's (RED FORCE) took off to bomb the same target. The formation was led by Major Fite, 828th CO. The forces rendezvoused, but not into wing formation. Thirty (30) P-51's escorted the formation from 1209 to 1330 hours. There were no early returns.

Twenty-one (21) aircraft (RED FORCE) were, over the primary target at 1232 hours and dropped 52.5 tons of 1000 lb. GP bombs from 17,900 feet, visually. Nineteen (19) aircraft (BLUE FORCE) were over the target at 1233 and 1246 hours and dropped 45.5 tons of 1000 lb. GP bombs from 20,000 feet. All, boxes scored hits within the central and eastern portions of the yard and the target was well hit. No enemy aircraft were seen and no flak encountered. There were no casualties or damage.

Weather: There were 57/10s low clouds at 3000 feet over the base at take off with 10+ miles visibility. There were 1-3/10s low clouds over the target.

Thirty-nine aircraft landed at base at 1328 hours. Aircraft 474 landed at Zara to refuel. Having feathered no. 3 engine approximately 50 miles west of Zara, the aircraft left at Zara and the crew returned to base aboard a 460th aircraft later.

In the middle of the afternoon, on the days we didn't fly, we listened for the hum of the returning planes. When the hum grew into a roar, everyone emerged from their tents. All eyes turned upward counting the returning planes to see how many were missing. Then we hitched a ride to the airstrip to greet the returning crews. The comradeship was evident between crews. We helped them to carry all the gear that had to be checked in after each mission. We were anxious to hear about the mission and helped them to talk about it. The Red Cross wagon was there with doughnuts and coffee. We discussed the mission while they were waiting their turn to go into critique. On the days we flew, they were there to greet us.

**MISSION 160, 19 MARCH 1945. TARGET: MUHLDORF MARSHALLING YARDS, GERMANY. BOMBS: 1000 LB GP, ALTITUDE: 20,000 FEET**

Muhldorf was an important railroad hub, with lines running to Munich, Regensburg, Linz, Salzburg and Rosenheim.

## MISSION NO. 161 - 20 MARCH 1945

At 0933, 20 B-24's, comprising BLUE FORCE and led by Capt. Jacob S. Disston III, 830th Ops Officer, and at 0949, 19 B-24's, comprising RED FORCE and led by Major Dolim, 829th Deputy CO, took off to bomb a primary target in Austria. Prevented from bombing the primary target, both forces bombed AMSTETTEN MARSHALLING YARD in Austria, the 1st alternate target. Wing rendezvous was not effected. Thirty P-38's escorted the forces from 1256 to 1430 hours. Aircraft 441 and 391 returned early.

At Fiume, a build up in the weather began. The formation managed to stay on top of the clouds over the Adriatic and at Judenburg an altitude of 22,500 feet was reached. The density of the clouds and their tops increased in altitude along the route ahead. Unable to climb above the clouds the leader of RED FORCE let down through them, having learned that the other Groups were at 20,000 feet. BLUE FORCE was somewhat slower in letting down, resulting in "B" box losing the formation. "B" box never rejoined the formation.

Slightly beyond Judenburgh, the lead Group gave notice that it was impossible to bomb the primary target by visual means because of the weather. It was decided at that time to bomb the 1st alternate target at Amstetten.

Seventeen aircraft of RED FORCE were over the target at 1327 hours and dropped 42 tons of 1,000 lb. GP's from 19,000 feet. Fourteen aircraft of BLUE FORCE were over the target at 1328 hours and dropped 35 tons of 1,000 lb. GP's from 17,500 feet. The bomb run was a combination of PFF and visual. Results: Thirty strikes in the west or main marshalling yard and 5 strikes in the loco depot area with 2 direct hits on the roundhouse. All tracks were cut in the east end of the yard and 25 to 50 units of rolling stock were destroyed or damaged. No flak was encountered over the target. SIH flak encountered on route from which 2 aircraft received minor damage.

Weather over the base was clear with 8 to 10/10 cirrus based at 20,000 feet to 22,000 feet. Thirty-seven aircraft landed at 1616 hours.

## MISSION NO. 162 - 21 MARCH 1945

At 0719 hours, 28 B-24's and 4 spares took off to bomb the NEUBURG A/D in Germany. The formation was led by Major Pruitt, 830th CO. Inline formation was effected. Forty (40) P-51's escorted the formation from 1044 to 1319 hours. One of the spare aircraft 834 returned to base. Difficulty was experienced in maintaining Wing formation. As the bomb run was started, it was noticed that the 47th BW was following very closely behind. The target was picked up without difficulty but in all the maneuvering at the IP, the formation became badly scattered, resulting in one Group being immediately above and one immediately below the 485th. A rally was made to the right with the Group endeavoring to line up for a second attempt on the target. This time the Group was hindered by the 47 BW formation. Six (6) aircraft of "C" box dropped on the first attempt, dropping 12.78, tons of clustered frags at 1156 hours from 21,700 feet. Photos revealed a good bomb pattern on the west end of the Airdrome, hitting the taxi strip, destroying two parked aircraft and damaging adjacent installations.

The formation circled from the PT, intending to attack the 1st alternate target. But before reaching it, it was learned that the area was closed in by clouds. Then a turn was made to the 2nd alternate target. The formation was denied the privilege of dropping its bombs. The bomb bay doors were opened and a bomb run started as some other Group passed beneath the formation just as the bombs were to be released. The leader realizing that some of his crews were beginning to worry about their fuel supply decided to abandon the mission and return to base. No enemy aircraft were seen and no flak encountered.

Weather: It was clear over the base at take off.

Twenty-five (25) aircraft landed at the base at 1506 hours. Aircraft 515, 729, 276, 419, 421 and 931 landed at Zara, refueled and retired to base later.

## MISSION NO. 163 - 22 MARCH 1945

At 0825 hours, 40 B-24's took off to bomb the HEILIGENSTADT MARSHALLING YARD at Vienna, Austria. The formation was led by Col. Cornett. Wing formation was accomplished and 30 P-38's escorted the Group from 1159 to 1340 hours. Aircraft 368 and 801 returned early. The bomb run was begun on PFF. Thirty-eight aircraft were over the target at 1246 hours and 36 aircraft dropped 70.25 tons of 100 lb. GP's from 23,100 feet. Photos showed scattered patterns on and around the target, with one pattern extending across the yard. A few strikes were made on the approaches to the bridge across the river.

No enemy aircraft were encountered. On return at 1353 hours when the formation was at 45°29N - 16°21E an aircraft was heard calling for the fighter escort, stating that enemy aircraft were attacking. The position was given as 15 miles north of Zara, Yugoslavia.

The formation was exposed to IAH flak over Vienna for 5 minutes. SAH flak was encountered on the outskirts of Bratislava. After bombs were away, the lead ship feathered no. 3 engine and the leader called for the Deputy Leader, Capt. Carl O. Bostrom, 831st Flt Cmdr, to take over lead, stating that his aircraft was badly hit and that he was going to attempt to get behind the Russian lines. His aircraft was list seen flying on a heading of 90 degrees at 48°32N - 17°27E. The leader of the fighter escort was heard dispatching fighters to escort a B-24 to Russia. The 1st attack unit rallied to the left and then turned south getting into the outer perimeter of the flak defenses of Bratislava. The 2nd attack unit rallied to the right.

Return was made without incident. Aircraft 416 landed at Zara with control cables damaged and wounded engineer.

At the base at take off time, the weather was good with only 3/10 stratocumulus at 12,000 feet. Over the target there were 5/10 middle clouds. Thirty-six aircraft landed at base at 1534 hours. Twenty-three aircraft were damaged by flak: 19 aircraft received minor flak damage; 4 aircraft received major flak damage: 779, 852, 038 and 507.

829th Bomb Sqdn Aircraft 50606

| | |
|---|---|
| P | Capt. Lloyd Allan |
| CP | Col. John B. Cornett |
| B | Maj. Howard M. Cherry |
| N | 1st Lt. Edward J. Anderson |
| N | 2nd Lt. Charles W. Carroll |
| ENG | TSgt. Don M. Hysell |
| RO | TSgt. John G. Mas |
| G | Sgt. Charles E. Spargur, Jr. |

G     SSgt. William Y. Sanders
N     1st Lt. Mentor Metaxas
G     SSgt. James A. Ridout

## MISSION NO. 164 - 23 MARCH 1945

At 0757 hours, 37 B-24's took off to bomb the GMUND MARSHALLING YARD in Austria. The 1st attack unit was led by Lt. Col. Cairns, Gp CO and the 2nd attack unit was led by 1st Lt. James C. Carlin, 829th Flt Cmdr. Line rendezvous was effected. Thirty (30) P-38's escorted the formation from 1106 to 1250 hours. Aircraft 929 returned early.

Thirty-six (36) aircraft were over the primary target at 1145 hours and dropped the bomb load from 20,000 feet. Eighteen (18) aircraft dropped 35.9 tons of 100 lb. GP bombs and 18 aircraft 36 tons of 250 lb. GP bombs, visually. Photos revealed the target was smoke obscured from previous bombing. All bombs, except from "B" box, which hit 800 feet to the north of the yard, were in the marshalling yard. No enemy aircraft were seen. No flak was was encountered, although one burst was seen at the primary target below and behind the formation. There were no damage or casualties.

Weather: On take off, it was clear over the base. It was clear on the target area.

Thirty-five (35) aircraft landed at 1456 hours and one aircraft landed at 1558 hours.

On return, a 4-engine aircraft was observed ditching 20 miles off the Italian Spur. BIG FENCE was notified and aircraft 859 was dispatched to circle the area until Air-Sea Rescue arrived. This aircraft located some debris in the water, believed to have been part of a tail section and part of a life raft. Aircraft 859 circled the area until a PBY arrived at 1430 hours.

## MISSION NO. 165 - 25 MARCH 1945

At 0812 hours, 37 B-24's took off to bomb the NEUBURG Airdrome in Germany. The 1st attack unit was led by Major Fite, 828th CO and the 2nd attack unit was led by Captain Carl O. Bostrom, 831st Flt Cmdr. The Group joined the Wing formation on course over the Adriatic. Thirty (30) P-38's escorted the formation from 1200 to 1426 hours. Aircraft 836 and 376 returned early.

Thirty-five (35) aircraft dropped 73.98 tons of frag bombs - 36 clustered - from 21,900 feet, visually. Results were very good. The main bomb pattern crossed the northwest dispersal area with some bombs extending over and into the Neuburg Station and sidings. Others were in the northeast dispersal area and to the northwest, extending across the railroad to Ingolstadt. No enemy aircraft were seen. No flak was encountered and there were no casualties or damage.

Weather: At take off, it was clear over the base. There were 1-3/10s cirrus clouds above 24,000 feet only.

Thirty-three (33) aircraft landed at 1604 hours. Aircraft 407 and 801 landed at a friendly field to refuel. Aircraft 407 returned later, while aircraft 801 stayed for repairs and the crew returned by aircraft from the 461st BG later.

## MISSION NO. 166 - 25 MARCH 1945

At 0804 hours, 32 B-24's took off to bomb the PRAGUE LETNANY AIRDROME in Czechoslovakia. The 1st attack unit was led by Major Dolim 829th Deputy CO and the 2nd attack unit was led by 1st Lt. R. E. Juday, 830th Flt, Cmdr. The Group assembled into wing formation. Thirty-five (35) P-51's escorted the formation from 1136 to 1413 hours. Aircraft 852 took off late and unable to locate the formation, made a visual bomb run on Udine Airdrome in northern Italy, dropping 2.16 tons of 36 clustered frags from 22,000 feet. Aircraft 829 and 414 returned early.

Thirty (30) aircraft were over the primary target at 1220 hours and 29 aircraft dropped 61.08, tons of 36 clustered frags from 22,300 feet, visually. Results: Excellent. A good bomb pattern was on the landing strip, buildings and installations adjacent. Smoke prevented a damage assessment of the installations or of the number of aircraft damaged.

Weather: There were a few high cirrus clouds over the base at take off. Also, a few high cirrus clouds over the target area. Over the base on return, there were 6/10s cirrus clouds.

No enemy aircraft were encountered. Two Me 262's were seen between the IP and the target. Both aircraft were low and at some distance and neither made any attempt to attack the formation. Five unidentified twin-engine aircraft were seen in the target area. SIH flak was encountered for two minutes. There was no damage or casualties. Aircraft 438 and 875 landed at Zara, refueled and returned to base.

Twenty-eight (28) aircraft landed at base at 1631 hours.

Prague/Letnany Airdrome, Prague. The flak gunners had the range but not the altitude. There was a heavy blanket of flak about 500 feet below us. LC Navigator.

## MISSION NO. 167 - 26 MARCH 1945

At 0905 hours, 26 B-24's took off to bomb the BRATISLAVA RANGER MARSHALLING YARD in Czechoslovakia. The 27th aircraft, 834, got stuck on the taxi strip, blocking traffic for several minutes and did not take off. At 1007 hours, 5 additional aircraft took off, 4 of which joined the formation after assembly. Aircraft 394, unable to catch the formation joined the 459th and bombed the Szombathely marshalling yard in Hungary. Wing rendezvous was accomplished. Thirty P-38's escorted the Group until 1330 hours. Aircraft 394 dropped 21 tons of 250 lb. GP's from 22,500 feet. Aircraft 694 returned early.

Twenty-nine aircraft were over the primary target at 1321 hours and 28 aircraft dropped 61 7/8 tons of 250 lb. GP's from 22,000 feet. Photos showed two bomb patterns in the marshalling yard. One on the car repair and shop area. The other on the eastern portion of the marshalling yard, probably cutting all tracks in that area and inflicting much damage to the rolling stock.

No enemy aircraft were seen. MAH flak was experienced over the primary target for 3 minutes. On the rally, weather was encountered which caused the formation to scattered. "B" box lost the formation and returned to base alone. This box encountered MAH flak at Moosbierbaum and at Nagykanizsa.

One aircraft is missing. Aircraft 930, flying no 3 position in "A" box was hit in no. 3 and no. 4 engines by flak over the target. Immediately after bombs away, this aircraft peeled off to the left, loosing altitude fast and went into a steep bank and when last seen, was heading east apparently under control. Later the formation heard the fighter escort reporting 9 chutes from a B-24 aircraft having lost 2 engines. Six aircraft landed at Zara for gas with only two returning later in the day: 534, 421, 819, 376, 528 and 361. The later two refueled and returned late.

At take off time there were 8-10/10 cirrus above 23,000 feet. It was cloudy over the base on return from the mission. Over the target there were cirrus clouds at 23,000 feet with 2-4/10 patchy middle clouds at 10-12,000 feet.

Twenty-three aircraft landed at 1712 hours. Two aircraft, 294 and 414, received major flak damage. Three aircraft received minor flak damage. There were no casualties.

830th Bomb Sqdn Aircraft 49930
P       1st Lt. George M. Manuel
CP      2nd Lt. Vernon P. Lovejoy
B       Capt. Victor Vergara
N       2nd Lt. Richard A. Labarron
ENG     T Sgt. Rolland E. Roller
RO      SSgt. Chester E. Konkolewski
G       Cpl. John J. Brennan, Jr.
G       Cpl. Raymond R. Haden
G       Cpl. Rupert E. Hazen
G       Cpl. George F. Raidel

Major Pruitt, Formation Leader: Two burst of flak exploded between the lead ship and 930, cutting control cables of the lead ship and hitting no. 3 and 4 engines of 930. Immediately after bombs away, 930 peeled off to the left loosing altitude fast and went into a. steep left turn, pulled out and headed east. Several minutes later, fighter escort reported seeing 9 chutes from a B-24 having 2 engines feathered.

## MISSION NO. 168 - 30 MARCH 1945

Four (4) B-24's took off to bomb the VIENNA NORTH STATION AND GOODS DEPOT in Austria by PFF. Aircraft 034, pilot 1st Lt. Herman Bosma. 831st off at 0800 hours, Aircraft 779, pilot 1st Lt. Thomas Peyton, 829th of at 0801 hours, Aircraft 517, pilot 1st Lt. August J. Horvath, 828th off at 0802 hours and Aircraft 758, pilot 1st Lt. Neil T. Priskow, 830th off at 0803 hours.

Aircraft 779 turned back at 1045 hours because of malfunctioning PFF equipment. Aircraft 034 arriving at the IP at 1115 hours, found that a large break in the clouds existed over Vienna which permitted the entire city to be identified visually. Having been briefed to attack the target only if it was overcast, the pilot decided to abandon the mission and return to base. At 1116 hours a turn was made to the north and the city of Vienna was circled to avoid flak defenses. A course was flown for the 1st alternate target which was also found to be clear. Aircraft 034 returned to base at 1356 hours.

Aircraft 517 arrived over the target area a few seconds later and found the same conditions. A course was flown to Graz and then to base, landing 1417 hours.

Aircraft 758 arrived over the target several minutes later and found that clouds had moved in over Vienna, sufficiently to provide ample protection to warrant a PFF bomb run. One ton of 500 lb. RIDX bombs were dropped at 1411 hours from 24,500 feet. Aircraft 758 landed at base at 1405 hours.

Weather: On take off, over the base there was an overcast at 11,000 feet with lower broken clouds at 3,000 feet. Over the target there were 8/10s riddle clouds with tops at 17,000 feet and rapidly clearing to the north. Temperature at 25,000 feet was -35 Degrees C/

No enemy aircraft were encountered. Aircraft 758 reported six unidentified aircraft near Zagreb, flying at 10,000 feet at 1205 hours. None of the aircraft encountered flak. There was no damage or casualties.

## MISSION NO. 169 - 31 MARCH 1945

At 0807 hours, 32 B-24's took off to bomb a primary target in Austria. Encountering adverse weather enroute, the VILLACH MARSHALLING YARD in Austria, in lieu of the primary target was attacked. The formation was led by Lt. Col. Atkinson. In line formation was accomplished. Twenty (20) P-51's escorted the formation from 1230 to 1335 hours. Aircraft 899, 067, 656, 438 and 368 returned early.

After completing the Wing rendezvous, the formation climbed on course without incident until encountering weather off Ancona. The formation flew in this weather for approximately a half-hour before being advised by the wing leader to circle to the left and climb above the layer. A total of six 360's turns were made before the formation broke through to the top at 22,000 feet at 1105 hours. The formation proceeded on course being approximately 47 minutes behind schedule. The leader then decided to select an alternate target suitable for either visual or PFF bombing. It was decided to bomb Villach marshalling Yard in Austria. Bombing by PFF and visual means, 27 aircraft were over the target at 1252 hours and 26 aircraft dropped 58.5 tons of 250 lb. GP bombs from 25,000 feet. Photos showed that the bombs fell along the Drava River south of the north marshalling yard causing no apparent military damage.

No enemy aircraft were seen. SAH flak was experienced over the target for one minute. One aircraft received minor flak damage. There were no casualties.

Weather: Over the base at take off, there were high cirrus clouds. Over the target there were 7/10s stratocumulus and cumulus clouds with tops at 15,000 feet. Over the base on return there were 6/10s cirrus and 2/10s cumulus at 3000 feet.

Twenty-five (25) aircraft landed at 1510 hours. Aircraft 419 landed at Zara and returned later. Aircraft 758 landed at Zara and remained with the crew flying to base in 460th aircraft.

## UNIT HISTORY, 485TH BG (H) 1 APRIL 1945 TO 30 APRIL 1945

During the month of April, the 485th BG flew 18 combat missions, making a grand total of 187 missions in the European Theater of Operations as of 25 April 1945.

On these missions, 822.35 tons of bombs were dropped. Results were very good, with the largest percentage of the bombing being visual. During the period of 10 May 1944 to 25 April 1945, 10,550.15 tons of bombs were dropped on enemy installations.

Five aircraft and their crews were lost during the month due to enemy action.

| | |
|---|---|
| Aircraft on hand - 1 April 1945 | 65 |
| Aircraft on hand - 30 April 1945 | 1 |
| Aircraft lost to operational accidents | 2 |
| Aircraft lost to non-operational accidents | 1 |
| Aircraft lost in combat | 5 |
| Aircraft transferred to 15th AF Service Command | 10 |
| Aircraft Gained | 13 |
| Aircraft transferred to Depot 52, Gioia | 56 |
| Aircraft transferred to 15th AFSC | 3 |

On 27 April 1945, warning orders were received preparatory to moving the unit to the U.S. for rehabilitation and conversation. During the latter part of the month rumors that the Group would move soon became prevalent. These rumors came to a climax on 24 April when the engineers began to remove the steel matting from the hard stands, for it was the day following that an official announcement was made that the group had completed its mission.

Then came a period of feverish haste with every section packing its records and equipment and with all personnel being processed. That rush continued until the unit entrained for the port of embarkation.

AWARDS

| | |
|---|---|
| DFC's | 33 |
| BRONZE STARS | 2 |
| PURPLE HEARTS | 2 |

## MISSION NO. 170 - 1 APRIL 1945

At 0919 hours, 32 B-24's took off to bomb a primary target in Austria. Lt. Col. Cairns led the 1st attack unit and 1st Lt. James W. Brady, 831st Flt. Cmdr, led the 2nd attack unit. Weather conditions at the primary target and alternate target areas prevented any of the targets from being attacked. The Group assembled into Wing formation. There was no fighter escort. Aircraft 727 and 694 returned early.

From Andria, a direct course was flown to Urgada, the formation arriving there as scheduled at 1330 hours at 15,000 feet. About 1.5 hours before target time, the weather aircraft reported thick cumulus from 22,000 to 27,000 feet in sector G, and clear weather below 24,000 feet in the target area. The 464th and 465th BG abandoned the mission because of weather conditions. One of these Groups reported the base of the weather at 19,000 feet and tops unknown. The Group leader continued enroute thinking that perhaps the primary target or an alternate target could be attacked. However, since the weather was moving in from the west, it was realized that all of the alternate targets were probably obscured.

A let down from 19,000 feet to 17,000 feet was made in an effort to get below the cloud deck. At this time the weather ahead was lower than 17,000 feet. Knowing that suitable bombing altitudes could not be reached if the formation was taken under the weather, the leader decided to abandon tie mission. The formation turned back at 1238 hours. No enemy aircraft were seen or no flak was encountered.

Thirty aircraft returned 75 tons of 500 lb. RDX's, landing at 1436 hours.

## MISSION NO. 171 - 2 APRIL 1945

At 1008 hours, 32 B-24's took off to bomb the GRAZ MAIN MARSHALLING YARD (PT) in Austria. The formation was led by Lt. Col. Andrus, Deputy Gp CO. Wing rendezvous was effected. Fighter escort was sporadic with P-51's coming and going at random from 1318 to 1435 hours. Six to nine fighters only were seen at a time. There were no early returns.

Thirty-two (32) aircraft were over the target at 1248 hours and 11 aircraft dropped 77.5 tons of 500 lb. RDX bombs from 21,400 feet, visually. Photos showed a good bomb pattern on the south end of the south yard, causing considerable damage to installations and rolling stock.

No enemy aircraft were seen. MAH flak was experienced over the primary target for two minutes. Aircraft 336 received major flak damage and seven other aircraft received minor flak damage. There were no casualties.

Thirty-one (31) aircraft landed at 1619 hours. Aircraft 409 landed at Zara, refueled and returned to base.

## MISSION NO. 172 - 5 APRIL 1945

At 0853 hours, 32 B-24's took off to bomb a primary target in Italy. Col. C. A. Clark, Jr., 55th Bomb Wing Executive Officer, flying with 1st Lt. Lloyd O. Simpson, 831st pilot, led the formation. The Group formed formation with the 460th BG. Twenty (20) P-38's escorted the formation from 1205 to 1230 hours. Aircraft 819 returned early as the prop governor on no. one engine malfunctioned, landed at 0929 hours.

The Group proceed alone to Castelgoffreda. The weather to this point had been good. A front could be seen lying over the target area to the east. A turn was made on the bomb run but the target was not visible. Three minutes later, the formation turned with the leader calling the other Groups, stating that he would fly the reciprocal course to the 1st turn point at which time he would give them his decision as to the target he would attack. It was decided to bomb the TURIN/LINGOTA LOGO DEPOT, a last resort target in Italy.

Thirty-one (31) aircraft were over the target at 1318 hours and 28 aircraft dropped 70 tons or 1000 lb. RDX bombs from 23,500 feet. Results: There were two direct hits on the turntable of the south round house and a number of strikes in the loco depot area. There were bomb strikes in the north end of the marshalling, yard, damaging rolling stock. The ball bearing factory east of the loco repair shop received several direct hits. On the highway bridge adjacent to the yard, two strikes were made on the approaches.

Weather: There were 9-10/10s stratocumulus clouds at 1000 feet over the base and visibility was 10 miles. In the target area there were 4/10s cirrus clouds at 25,000 feet. On return there were 5/10s stratocumulus clouds at 3000 feet over the base. No enemy aircraft were seen, no flak encountered and no casualties or damage. Thirty-one (31) aircraft landed at 1610 hours.

At 0911 hours, 6 B-24's took off to bomb a primary target in Italy. The formation was led by Lt. Col. Cairns. Wave formation was not accomplished. Thirty (30) P-51's escorted the formation from 1215 to 1300 hours. Aircraft 038 and 724 returned early because of bad weather, the formation did not drop their bomb load. No enemy aircraft were seen. No flak was encountered. There were no casualties or damage.

Weather: Over the base on take off there were 9-10/10s stratocumulus clouds at 1,000 feet, and visibility was 10 miles. In the Verona, Italy area there were 10/10s cirrus clouds based at 18,000 feet.

Four (4) aircraft returned 10 tons of 100 GP bombs it 1445 hours.

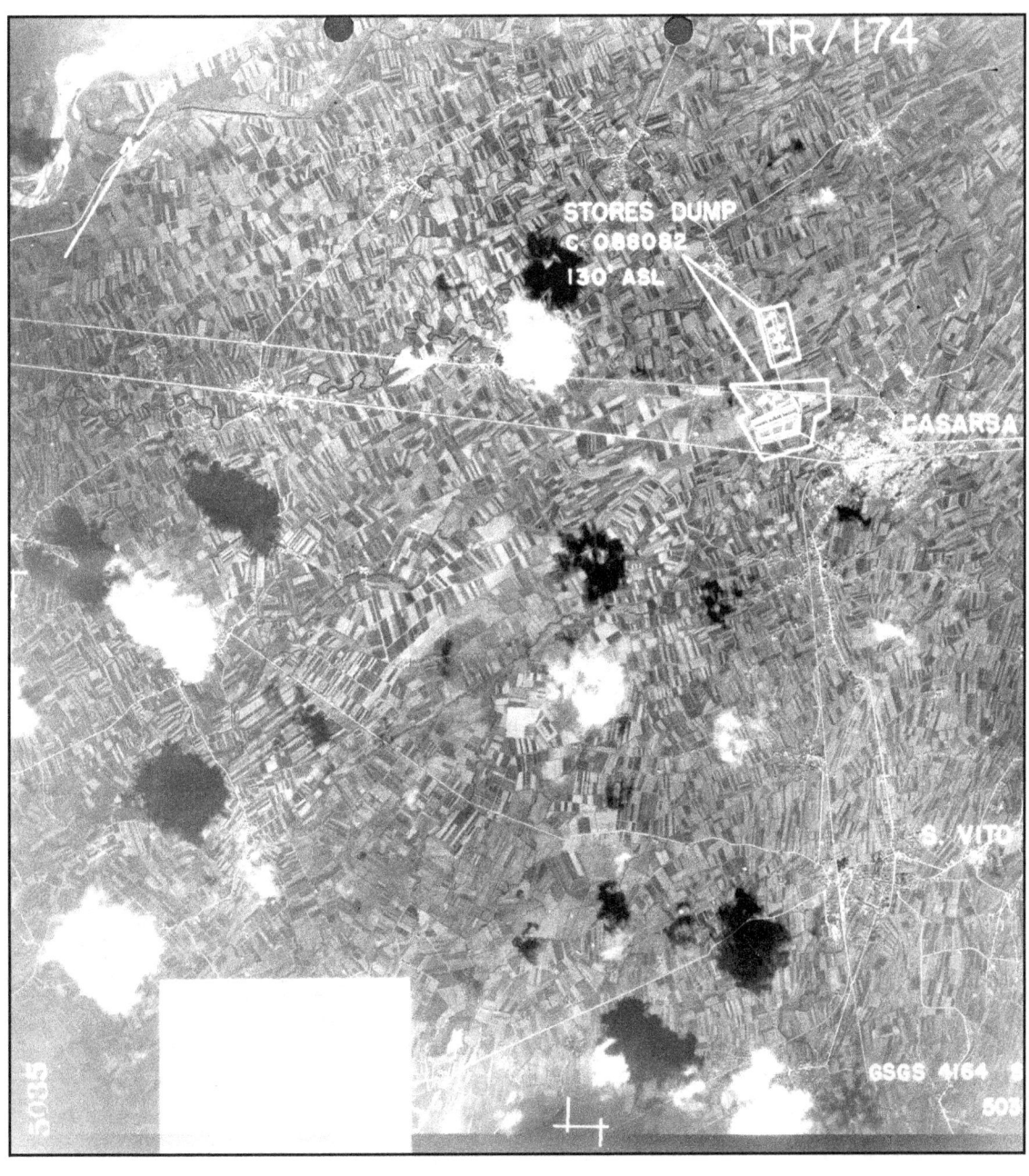

**CASARSA ITALY, 4 APRIL 1945. STORES DUMP. RECONNAISSANCE PHOTOGRAPH**

On April 23, 1945 the 485th bombed the Casarsa Diversion Bridge (not shown) during "Operation Strangle", in an effort to cut off the retreating German army.

## MISSION NO. 173 - 7 APRIL 1945

At 1051 hours, 31 B-24's took off to bomb a primary target in northern Italy. Major Fite was scheduled to lead the formation, but because of illness during assembly, Col. Fite returned to base and Captain Wm. R. Fritz, 828th Flt Cmdr took over the lead. Aircraft 410, the lead aircraft returned to base discharging Major Fite and then returned to the formation at 1147 hours, flying in no. 2 position. Wing rendezvous was effected. Fifteen (15) P-51's escorted the formation from 1430 to 1525 hours. Aircraft 394 took off late, at 1141 hours. Unable to locate the formation, 394 returned to base. An attempt was made without success to intercept the 465th and the 464th BG's. The aircraft turned back at 1351 hours and landed at 1529 hours.

Opposite Sibenik, Cirrus clouds were encountered. The formation climbed above the clouds to 18,000 feet and continued on course. The cirrus clouds continued with tops at 23,000 feet at Mooslbacher. The course was continued until the 304th BW was observed turning back. At this time the Wing leader was contacted and notice was received that the mission was being abandoned. At this point, a complete undercast existed with a front extending above 25,000 feet, lying between the formation and the primary target and the alternate target areas. During the movement, "D" box became separated and returned to the base alone.

Thirty (30) aircraft landed at 1717 hours. Aircraft 410 lost an engine and returned slightly ahead of the formation. Bomb load - 77.5 tons of 1000 lb. RDX bombs. No enemy aircraft were seen, no flak encountered and there were no casualties or damage.

## MISSION NO. 174 - 8 APRIL 1945

At 0828 hours, 30 B-24's took of to bomb the FORTEZZA MARSHALLING YARD in northern Italy. The formation was led by Major Henry P. Dolim, 829th CO. The Group rendezvous with the 460th BG. Twenty (20) P-51's escorted the formation from1145 to 1257 hours. Aircraft 819, 440, 834, 694 and 407 returned early.

Bombing results were very poor. No enemy aircraft were seen. MAH flak was encountered for five to six minutes over the primary target. Eight aircraft received minor damage. Two aircraft received major flak damage. Sgt. Frank A. Syabla, gunner on aircraft 859 and Captain Rusiewicz, Group Bombardier, lead bombardier on aircraft 714 were wounded.

Weather: At take of, there was a high overcast at 12,000 feet over the base. Over the target there were a few cumulus clouds and visibility was unrestricted.

Twenty-three (23) aircraft landed at 1538 hours. Aircraft 714 landed at Zara and returned later. Aircraft 859 remained at Zara for repairs and aircraft 875 remained at Bari, Italy.

Fortezza Marshalling Yards are in the Brenner Valley. Our bomb run took us across the valley to minimize the time in the flak zone. Another group was heading for the target at the same time we were. They must have outranked us because we yielded and completed a 360 degree turn and followed them over the target. By now the flak gunners had the range figured out. As we started over the target, flak exploded beneath our plane which gave Boatright, the ball gunner, a jolt. The old bird raised up, shook and wobbled and settled down. I thought the valley was narrow, but it seemed like it took forever to get across. We ended up with a bunch of flak holes, but were spared again. LC, Navigator.

## MISSION NO. 175 & 176 - 9th & 10th APRIL 1945

At 1137 hours, 42 B-24's took off to bomb tactical targets in Italy. Lt. Col. John E. Atkinson, Deputy Gp CO led the RED FORCE and Major Joseph Gill, 828 Ops Officer, led the BLUE FORCE. Wing Rendezvous was effected. There was no fighter escort as the fighters provided general coverage for the entire target area. There were no early returns. Bombing was visual.

Twenty-one aircraft of RED FORCE were over the primary target at 1435 hours and dropped 43.2 tons of 36 clustered fragmentation bombs from 20,600 feet. Results: Excellent concentration of bombs covering the MPI. Twenty-one aircraft of BLUE FORCE were over the primary target at 1500 hours and dropped 44.64 tons of 36 clustered fragmentation bombs . No bombs fell short of the bomb release line.

No enemy aircraft were seen. SIH flak was encountered after bombs away on the rally for approximately 1 minute by each force. There were no casualties or battle damage.

The weather at the base was 8/10 cumulus at 2500 feet with visibility unrestricted. The target area was clear.

Forty-two aircraft landed at 1708 hours.

At the briefing, we learned that the British 8th Army was going to try to break through the heavily defended front start chasing the Germans up the "Boot". We were scheduled to lay down an aerial barrage from the Adriatic. There was nothing to compare to an American barrage). The Senio River seperated the two armies and our ssignment was to drop frags just across the river. Our main concern was was to make sure that no bombs fell on the wrong side. To help identify the Front, our troops were to lay down sheets of white cloth in the shape of large "T's." Also a radio beam was shot parallel to the Front. Before dropping our bombs we to identify the Front three ways; the River, the radio beams and the white "T's". Also we were instructed to open the bomb bay doors over the Sea before the IP.

We had no trouble finding the Front; that's where all the flak was. We crossed the river, dropped the frags and were ready to turn back across the lines. Just then we saw a B-24, from another group, heading towards us with flames coming out of the #3 Engine and the cockpit. The plane passed under us and blew up. The gunners reported seeing only one chute. Lynn Cotterman, Navigator.

## MISSION NO. 177 - 11 APRIL 1945

At 0751 hours, 31 B-241s, led by Lt. Col. Douglas M. Cairns, took off to bomb the CAMPO DI TRENS RAILROAD BRIDGE in Italy. Rendezvous with the 460th BG was effected. Twenty P-51's escorted from 1050 to 1318 hours. There were no early returns.

Thirty-one aircraft were over the primary target at 1207 hours and dropped 77.5 tons of 500 lb. RDX's from 23,500 feet, visually. Several hits were scored on the bridge. No enemy aircraft were seen. MAH flak was encountered, directed principally at the last two boxes, for more than 2 minutes.

Aircraft 95517 was hit over the target and dropped out of formation, losing altitude. The aircraft continued down in a slow spiral and finally crashed at 1210 hours. Eight chutes were definitely seen to have opened soon after the aircraft was hit. One crew reported two other chutes opening a few seconds before the crash. The aircraft crashed approximately 10 miles southwest of the target.

Aircraft 50421 radioed that the pilot and co-pilot were both wounded and that no. 3 engine was feathered and that the aircraft was low on gas. Aircraft requested a heading to the nearest landing strip and was directed to Falconara. Aircraft was last heard from at 1310 hours at which time it was south of the front lines.

In addition, 6 aircraft received major flak damage: 740, 536, 727, 465, 424 and 908. Nine aircraft received minor damage.

At take off time, the weather was clear over the base. In the target area there were a few cirrus at 25,000 feet with unrestricted visibility.

Twenty-seven aircraft landed at 1505 hours. Aircraft 441 landed at Falconara and 834 landed at Madna for gas and returned later.

830th Bomb Sqdn Aircraft 95517   830th Bomb Sqdn Aircraft 50421

| | | | | |
|---|---|---|---|---|
| P | 2nd Lt. Neil V. Rodreick | | P | 1st Lt. Donald L. Adams |
| CP | 2nd Lt. George A. Zukowski | | CP | 2nd Lt. Vincent J. Barison |
| B | 2nd Lt. Edward J. Reifer | | N | 2nd Lt. Armand G. Gazaille |
| RO | Sgt. Albert L. Scott | | ENG | Sgt. Joseph N. Catoggio |
| ENG | SSgt. Byron E. Taegue | | RO | Sgt. Frank D. Salters |
| G | Cpl. Edward R. Kaus | | G | Cpl. Thomas O. Moore |
| G | Cpl. Walter F. Ritz | | G | Cpl. Thomas F. Falcone |
| G | PFC Jean A. Robertson | | G | Cpl. John J. Chess |
| | | | G | Sgt. Marvin Nicholson, Jr. |

## MISSION NO. 178 - 12 APRIL 1945

At 0759 hours, 28 B-24's took off to bomb the PONTE DI PIAVE RAILROAD BRIDGE in northern Italy. The formation was led by Major Fite, 828th CO. Wing rendezvous was not effected. Fifteen (15) P-51's escorted the formation from 1146 to 1220 hours. Aircraft 059, losing an engine, returned early.

Twenty-seven aircraft were over the primary target at 1211 hours and 26 aircraft dropped 52 tons of 500 lb. RDX bombs from 21,000 feet. Results: Two or three direct hits and two or three hits on the south approach to the bridge. No enemy aircraft were seen, no flak encountered and there were no casualties or battle damage.

Weather: Over the base at take off there were 6/10s cirrus clouds and visibility was unrestricted. Over the target area there were a few cumulus clouds and 7/10s cirrus clouds above 24,000 feet. Visibility was 25+ miles.

Twenty-seven aircraft landed at 1418 hours.

## MISSION NO. 179 - 14 APRIL 1945

At 0811 hours, 31 B-24's took off to bomb the, OSSOPO M/T DEPOT in northern Italy. The formation was led by Major Henry P. Dolim, 829th CO. Wing rendezvous was not effected. Thirty (30) P-38's escorted the formation from 1100 to 1242 hrs. There were no early returns.

Thirty-one (31) aircraft were over the primary target at 1147 hours and 23 aircraft ("A", "B" and "D" boxes) dropped 57.5 tons of 500 lb. RDX bombs from 19,700 feet. Charlie box failed to drop their bombs and rallied to the left from the primary target and turned at Aviano for another bomb run. Unable to identify the target because of clouds, Charlie's box continued to Klagenfurt. There, the target was identified but insufficient time for the bombardier, they did not release the bombs. Charlie's Box made a second bomb run and 8 aircraft dropped 20 tons of 500 lb. RDX bombs on KLAGENFURT NORTH MARSHALLING YARD, visually. Results: Six direct hits were obtained in the Klagenfurt marshalling yard. Results of the bombing of the Ossopo Depot were poor. Bomb strikes extended from the south edge of the runway on the airdrome to the eastern section of the barracks area. No enemy aircraft were seen and no flak was encountered. There was no casualties or battle damage.

Weather: Over the base there were 2/10s middle clouds at 10,000 feet. Over the target there were 2-4/10s low clouds with a haze aloft.

Thirty-one (31) aircraft landed at 1352 and 1453 hours.

## MISSION NO. 180 - 15 APRIL 1945
BLUE FORCE

At 0905 hours, 11 B-24's took off to bomb a COMMUNICATIONS TARGET in northern Italy. The formation was led by Major William D. Ceely, 831st CO. The Group formed with the 460th BG. Twenty (20) P-51's escorted the formation from 1225 to 1348 hours. Aircraft 727 returned early.

At Muda, weather conditions were determined to be such as to prevent visual bombing of the primary target, 1st alternate target or the 2nd alternate target. A layer of cumulus clouds, 7-8/10s at 8,000 feet, covered the entire target area and extended west. The leader turned north at Muda, hoping to bomb Klagenfurt marshalling yard by PFF. 9/10s cumulus clouds at 12,000 feet covered the target and the Mickey Operator was unable to identify the target by PFF, after two attempts. The leader decided to abandon the mission and return to base. No enemy aircraft were seen and no flak was encountered. There were no casualties. Aircraft 138, 899, 638 and 931 landed at Zara, refueled, and returned to base after 1700 hours.

Weather: On take off over the base there were 10/10s cumulus clouds from 2000 to 6000 feet, with 10 miles visibility. In the target area there were 8-10/10s cumulus clouds with tops at 15,000 feet. Over the base on return there were 3/10s cumulus clouds from 2500 to 9000 feet with unrestricted visibility.

Six (6) aircraft landed at base at 1637 hours. (Bombs - 500 lb. RDX)
RED FORCE

At 1015 hours, 40 B-24's, led Major Fite, 828 CO, took off to bomb GUN INSTALLATIONS in the Bologna area, Italy. Wing formation was effected. Thirty P-38's furnished target and withdrawal cover, starting at 1402 hours. Aircraft 532 returned early.

Thirty-nine aircraft were over the primary target at 1407 hours and dropped 77.5 tons of 100 lb. GP's from 22,000 feet. Several strikes were scored in the target area, Savana River, and adjacent railroad damaged. Military damaged was undetermined.

No enemy aircraft were seen. Very little flak was encountered. A barrage of flak was seen ahead of the preceding Groups, with only a few burst approaching the vicinity of the 485th just before the bombs were released. Some of the bursts were red in color.

Aircraft 49024 is missing. Very little information is available as to where this ship was last seen. One crew reported having a call from this aircraft, stating that it was going to a friendly airfield. It is believed that the aircraft is possibly at Corsica.

Weather over the base was 10/10 cumulus from 2,000 to 6,000 feet with visibility of 8 miles. The target area was clear. On return from the mission, there were 3/10 clouds and visibility unrestricted at the base.

Thirty-eight aircraft landed at 1648 hours.

## MISSION NO. 181 - 16 APRIL 1945

At 0930 hours, 41 B-24's took off to bomb the GUN INSTALLATIONS in the BOLOGNA area in northern Italy. The formation was led by Lt. Col. Douglas M. Cairns, Gp CO. The Group formed into Wing formation. Several P-51's were observed in the target area, providing general fighter coverage. There were no early returns, although aircraft 801 turned back at the IP and landed at Lesi.

Forty (40) aircraft were over the primary target at 1321 hours and 21 aircraft dropped 45.62 tons of 250 lb. GP Bombs from 21,000 feet. The bomb strike was undetermined because of cloud coverage as shown on photos. The 2nd attack unit did not release their bombs because of failure to positively identify the target. No enemy aircraft were seen and no flak was encountered. There was no casualties or damage.

Weather: At the base on take off there were 5/10s cirrus and 2/10S low clouds. Over the target area there were 8-10/10s stratus and cumulus clouds at 8000 feet.

Forty (40) aircraft landed at base at 1609 hours.

## MISSION NO. 182 - 17 APRIL 1945

At 1024 hours, 41 B-241s, led by Major Henry P. Dolim, 829 CO, took off to bomb GUN INSTALLATIONS in the Bologna area in Italy. Wing formation was effected. P-51's provided fighter coverage over the target and were first observed at 1353 hours. There were no early returns. Forty-one aircraft were over the primary target at 1404 hours and 39 aircraft dropped 87.37 tons of 250 lb. GP's from 20,000 feet. Photos showed a concentration of bombs in the target area. No enemy aircraft were seen. No flak was encountered, however, several bursts of flak were observed about 1,000 yards ahead of the formation on rally. There was no damage or casualties.

The weather was clear over the base and the target.

Forty-one aircraft landed at 1636 hours.

## MISSION NO. 183 - 19 APRIL 1945

At 0748 hours, 42 B-241s took off to bomb a primary target in Austria. The formation was led by Lt. Col. Douglas L. Cairns, Gp CO. Wing formation was effected. Twenty (20) P-51's escorted the formation from 1137 to 1330 hours. Aircraft 728 returned early.

Forty (40) aircraft were over ROSENHEIM MARSHALLING YARD at 1255 hours and dropped 99.5 tons of 500 REX bombs from 21,300 feet. Aircraft 049 jettisoned at 1036 in order to keep in formation. Bombing was by visual means. Results: Photos showed a good bomb pattern of bombs covering the central and the south-east ends of the yards, damaging 50 to 100 pieces of rolling stock. Highway bridge south of the yard was destroyed and the installations adjacent to the southeast choke point were damaged. No enemy aircraft were seen. The formation did not encounter any flak. Aircraft 638 experienced MAH flak at Udine for 1.5 minutes, receiving minor flak damage. There were no casualties.

Aircraft 638 lost an engine and left the formation at Hallstait and attacked the airdrome at Udine at 1158 hours, dropping 2.5 tons of bombs from 18,500 feet. Bombs were observed to have hit slightly northwest of the hangar installations.

Several aircraft landed at friendly fields. Aircraft 638 landed at Falconara with an engine out. The aircraft was left there and the crew returned to base. Aircraft 414 landed at Zara, refueled and returned to base. Aircraft 908, 528, 834, 931 and 829 landed at Lesi, refueled and returned to base. Aircraft 407 landed at Forli, refueled and returned to base.

Weather: There were a few high thin cirrus clouds over the base. Over the target there were 5/10S cumulus clouds with tops at 13,000 feet. At the base on return there were 6/10s cumulus clouds with light rain.

Thirty-three (33) aircraft landed at base at 1531 hours.

## MISSION NO. 184 - 20 APRIL 1945

At 0833 hours, 42 B-241s, led by Capt. Harold F. Cline, Asst. GP Ops Officer, took off to bomb the GARZARA ROAD BRIDGE in northern Italy. Assembly with the 460th BG was effected. Twelve P-51's escorted the formation from 1150 to 1245 hours. Aircraft 656, which blew a cylinder on no. 4 engine, returned early.

Forty-one aircraft were over the primary target at 1220 hours and dropped 101.5 tons of 1,000 lb. RDX's from 21,500 feet. Photos showed center span had been cut by previous bombing. This Group damaged both approaches and blasted the road for some distance on the south side. Several near misses and probable hits on the structure. No enemy aircraft were seen. No flak was encountered and there was no damage or casualties.

At take off the weather was clear over the base and over the target there were high cirrus clouds.

Forty-one aircraft landed at 1437 hours.

## 21 APRIL 1945 (Mission Aborted)

At 0941 hours, 42 B-24's took off to bomb a primary target in northern Italy. The formation was led by Major Calvin Fite, 828th CO. The Group formed into Wing formation. There was no fighter escort. Aircraft 875, 507 and 049 returned early.

The formation arrived at the KP on schedule, having climbed through some weather during which the 460th BG became separated. The formation reached an altitude of 17,500 feet above a complete undercast. At 13,500 feet the leader decided to abandon the mission since at that tine the formation would have to lower to 8 to 10,000 feet in order to get beneath the weather. Rather than risk taking the formation through the weather or to chance going below it at 8,000 feet, the leader turned back at 1305 hours.

Weather: There were 1/10s cirrus above 20,000 feet, 10/10s thick middle clouds with bases at 2,000 feet at Siena. Low and middle clouds converged to the north, with the target area overcasted.

No enemy aircraft were seen, no flak encountered and there were no casualties or damage.

Thirty-nine (39) aircraft !ended at base at 1504 hours with their bomb load.

## MISSION NO. 185 - 23 APRIL 1945
### 1ST ATTACK UNIT

At 0757 hours, 20 B-241s, led by Major Dolim, 829th CO, took off to bomb the PADUA ROAD BRIDGE (Primary Target) in northern Italy. Wing rendezvous was not effected. Ten P-51's were observed over the target area furnishing general target area coverage. There were no early returns.

Nineteen aircraft were over the target at 1227 hours and 15 aircraft dropped 37.5 tons of 1,000 lb. RDX's from 23,000 feet, visually. Aircraft 834 returned to base and aircraft 819 jettisoned in the Udine area. Aircraft 740, 899 and 409 also jettisoned their bombs. Photos showed that both approaches for a distance of 1,000 feet were well covered with 2 direct hits on the west abutment.

No enemy aircraft were seen. SIH flak was experienced for .5 minute as the formation rallied from the target. First, the burst were high, then low and finally at the sane level as the formation, but trailing by several yards. There were no casualties or battle damage.

Aircraft 50819 is missing. It left the formation at 1158 hours with two engines out. The 325th FG reported that this aircraft was over enemy territory at 5,000 feet with 2 engines out and that it was doubtful if the aircraft could make friendly territory.

The weather over the base was 6/10 high clouds above 25,000 feet with 15 mile visibility. Over the target area it was clear.

Return was made without incident, other than several aircraft landing with feathered props. Eighteen aircraft landed at 1412 hours. Aircraft landed later.

### 2ND ATTACK UNIT

At 0812 hours, 21 B-241s, led by Major Fite, 828th CO, took off to bomb the CAVARZERA ROAD BRIDGE in northern Italy. Wing formation was not effected. Ten P-38's provided cover in the target area. Aircraft 474 and 724 returned early.

Nineteen aircraft were over the target at 1210 hours and dropped 47.5 tons of 1,000 lb. RDX's from 22,000 feet, visually. Photos showed bomb pattern covered north approach with one strike on north abutment and pattern south and right of the bridge. SIH flak was observed beginning approximately .5 minutes after bombs away and continuing for 2 minutes. The flak was both high and low and to the rear of the formation. No enemy aircraft were seen. There was no battle damage or casualties.

The weather was the same as for the 1st attack unit.

Nineteen aircraft landed at 1346 hours.

831st Bomb Sqdn Aircraft 50819
P    1st Lt. Nicholas J. DeMoso
CP   F/O James F. Henderson
N    F/O Joseph T. Harmuth
B    2nd Lt. Edwin S. Nichols
ENG  Cpl. Richard. K. Fitzpatrick
RO   Sgt. Vincent J. Noce
G    Cpl. William H. Majors
G    Cpl. Joseph A. Wachter
G    Cpl. Harold D. McDonough
G    Cpl. Arthur F. Schlagel

Aircraft 50819 called at 1158 hours, stating that it was losing altitude fast with one engine feathered. Location 46-16N - 12-30E

## MISSION NO. 186 - 24 APRIL 1945

At 0716 hours, 40 B-241s, led by Major William Ceely, 831st CO, took off to bomb the CASARSA DIVERSION BRIDGE in northern Italy. Rendezvous with the 460th BG was scheduled, however, was not accomplished as the Group was unable to locate the 460th. Eighteen P-38's were observed at 1030 hours which provided general cover. Aircraft 532 returned early.

Thirty-nine aircraft were over the target and dropped 96.75 tons of 500 lb. RDX's from 23,000 feet, visually. Photos showed the west half of the bridge covered with strikes, direct hits and near misses. West approach received several direct hits and near misses with one pattern of bombs over and to the southeast of the target. No enemy aircraft were seen and no flak was encountered. There was no battle damage or casualties.

Aircraft 724 left the formation after bombs away with no. 3 engine feathered and landed at Zara. The aircraft was left there for engine change and the crew returned to base by C-47 at 1625 hours. -

The weather over the base at take off was 1/10 cirrus at 24,000 feet and visibility was more than 15 miles, and on return there were 5/10 cumulus at 4 - 6,000 feet. Over the target there was 1/10 cumulus with tops at 10,000 feet

Thirty-seven aircraft landed at 1330 hours.

## MISSION NO. 187 - 25 APRIL 1945

At 0839 hours, 30 B-24's took off to bomb the LINZ NORTH MARSHALLING YARD in Austria. The formation was led by Lt. Col. Cairns, Gp CO. A line rendezvous was effected. Aircraft 694, 038, 596 and 385 returned early.

Twenty-six (26) aircraft were over the target at 1324 hours and dropped 65 tons of 500 lb. RDX bombs from 24,500 feet, visually. Photos showed a good bomb pattern covering assigned portions of the yard, damaging rolling stock and hitting loco depots, turntables and the highway overpass. A dog fight between a P-38 and an unidentified aircraft was observed on the rally. IAH flak was encountered for three minutes. Aircraft 614 and 728 incurred major flak damage and 10 other aircraft received minor flak damage. SSgt Walker, tail turret gunner on aircraft 728 was wounded. Aircraft 409 landed at Zara, refueled and returned later to base. Aircraft 728 landed at Zara because of battle damage and for treatment of SSgt Walker.

Aircraft 414 is missing. This aircraft was hit by flak on the bomb run and dropped its bombs a few seconds early. It rallied away from the formation with no. 3 and 4 engines smoking, and headed toward Zara. Later, it was observed lagging behind the formation with fighter escort.

Weather: It was clear over the base on take off and clear in the target area. On return there were a few cumulus clouds over the base.

Twenty-three (23) aircraft landed at 1610 hours.

830th Bomb Sqdn Aircraft 50414

| | | | |
|---|---|---|---|
| P | 1st Lt. Joseph W. Cathey | G | Sgt. John E. Chamberlain |
| CP | 2nd Lt. Everett D. Banker | G | Sgt. Raymond F. Liebold |
| B | 1st Lt. Thomas S. Morrison | G | Sgt. Aldo Grandoni |
| RO | Sgt. Monte B. White | G | Sgt. Leon R. Wilins |
| ENG | Sgt. Charles E. Dyke | | |

**SECRET**　　　　　　　　**PILOTS FLIMSY for 25 APRIL 1945**　　　　　　　　**SECRET**

| | | | | | | |
|---|---|---|---|---|---|---|
| Crew Inspection | : | 0720 | Stations: | 0730 | Start Engines: | 0740 |
| Taxi | : | 0750 | Take off: | 0800 | | |

GROUP ASSEMBLY : Over home field at 8,000'
WAVE RENDEZVOUS : 465th leads line rendezvous over ANDRIA at 0917 and over 4200 1620 at 0937.
ORDER of FLIGHT : 304th, 49th, 47th, 55th in very close column (465-464-485-460), 5th wings.
FIGHTER ESCORT : 36 P-38's of 82nd fighter group will intercept 55th Wings at 4718 1251 at 1200B and provide escort on P.T.W.
KEY POINT : LEDENICE (4509 1451) at 1058 at 14,000'
START CLIMB to Bombing altitude at 4200 1620
PRIMARY TARGET : LINZ NORTH MAIN M/Y (Visually) at 1250
INITIAL POINT : WEGSHEID (4836 1347)
AXIS of ATTACK : 133° from IP - 150° from ST JOHANN
Method of ATTACK : By attack units
TARGET ELEVATION : 849'
BOMBING ALTITUDE : 1st attack unit - 24,5000'     465th - 22,500'
　　　　　　　　　　　2nd attack unit - 24,200'      464th - 23,500'
　　　　　　　　　　　　　　　　　　　　　　　　　460th - 25,500'

RALLY : Left to KRUEZEN (4819 1448)
FIRST ALTERNATE : AF #4     PASSAU M/Y (Visually only)
INITIAL POINT : DEGGENDORF (4850 1258)
AXIS OF ATTACK : By attack units.
RALLY : Right to ZELL (4819 1338)
ROUTE BACK : ZELL to ST LORENZEN to same as Primary.
SECOND ALTERNATE : AF #6     FRIELASSING M/T DEPOT (Visually only)
INITIAL POINT : TACHERTING (4805 1234)
AXIS OF ATTACK : 132°
METHOD OF ATTACK: By attack units.
RALLY : Right to UNKEN (4739 1243)
ROUTE BACK : UNKEN to TARVISIO then reverse route out.

THIRD ALTERNATE : AF #7 CORTINA AMMO DUMP (Visual Only)
INITIAL POINT : FORNI D.S. (4626 1235)
AXIS OF ATTACK : 306°
METHOD of ATTACK : By attack units.
RALLY : Left to PIEVE D.L. (4-29 1157)
ROUTE BACK : PIEVE to ARTEGMA (4615 1309) to POSTINA (4547 1413 to LEDENICE to Base.
SPECIAL INSTRUCTIONS:
　　　　　　　　　　　GAS LOAD:     2,700 gallons
　　　　　　　　　　　BOMB LOAD:   10 x 500# RDX fused .1 nose and .025 tail
Any A/C bringing in seriously wounded land at BARI or FOGGIA MAIN
Any A/C bringing in slightly wounded land at IESI or PANTANELLA
　　　　　　　　　　　5TH Wing attacks LINZ M/Y at 1330 to 1345
　　　　　　　　　　　47th Wing attacks LINZ M/Y at 1218 to 1223
　　　　　　　　　　　49th Wing attacks LINZ M/Y at 1210 to 1218
　　　　　　　　　　　304th Wing attacks LINZ M/Y at 1130 to 1140
Use evasive action point ST JOHANN only if bomb run is visual.
PFF for navigation only.

**SECRET**　　　　　　　　　　　　　　　　　　　　　　　　　　　　　　**SECRET**

**SECRET**　　　　　**FORMATION POSITIONS FOR MISSION 25 APRIL 1945**　　　　　**SECRET**

```
                         COL. CAIRNS
                         MAJ COLLINS
                         WY      614      P/F LT PHILLIPS
          MURPHY                             HICKS    *
          WF    728                         BU      034        P/F LT VELENTINE (823)
                                                               C/S - SOONUP
                         CATHEY
                         WK      414
          SCHUELER                           KING
          WI    614                         WL      296
                         BACON
                         WV      908


                                                               HAZEL
                                                               RN      875
                                            DEOBALD                          HARVATH   *
                                            RC     440                       RR     351
    c/s - SPRINGGREEN                                          ??????
                                                               RA      515
                                            DEEGAN                           ?????
                                            RI     929                       RE     052
                                                               BROWN
                                                               RS      038


                         PARSEGIAN
                         BE     596        P/F LT SCHIANSKY
          O'ROURKE                           SMITH    *
          YW    336                         YV      419
                         PEYTON                                C/S - VETREG
                         YI     740
          SCHILEY                            MARTIN
          YX    365                         YL      592
                         HEIMS
                         YS     067
          SCHILL                             HARKENRIDER
          WN    504      (830)              YB      416


                                                               JACKSON
                                                               BM      914
                                            ROSWALD                          LUDLOW    *
                                            BM     727                       BF     424
    C/S - RUBDOWN                                              BAKER
                                                               BC      834
                                            MC CARTHY                        CHAPLAN
    GROUND SPARES: YP - 517 PFF (829)       BB     829                       BO     138
                   WP - 915 N. BS (830)                        WARDEN
                                                               BS      409
                                            MURRAY                           HESS
                                            RO     528  (828)                WQ  536 (830)
```

COMMUNICATIONS:
  INTERPLANE    : VHF "B" & #1 command  465th Gp  BURGLAR 1
  BETWEEN GRUPS  : VHF "C"        464th Gp  BURGLAR 2
  WING TO WINGS  : VHF "A"        485th Gp  BURGLAR 3
  BOMBER TO FIGHTER: VHF "A"        460th Gp  BURGLAR 4
  AIR SEA RESCUE  : VHF "D"        5th Wg   FOOTHOLD
  EMERGENCY   : VHF "D"        47th Wg  PULPWOOD
  BIG FENCE    : VHF "A"        49th Wg  CHANNEL
  RECALL WORD  : POKER         304th Wg  RELLBUOY
FIGHTER C/S: LACEWORK  Wx A/C - OUT ENCORE 1
              BACK ENCORE 2
THE CALL DURING RENDEZVOUS: UPHILL 460th - SQANDER 465th - REWARD
                 464th - WINDBLOW 485th - THORAX

## 485TH BOMB GROUP MISSION RECORD

### Description of Abbreviations Used

| | | | |
|---|---|---|---|
| A/CF | Aircraft Factory | I/A | Industrial Area |
| Acft | Aircraft | KIA | Killed in Action |
| A/D | Air Drome | L/D | Locomotive Depot |
| AM/D | Ammunition Depot | MIA | Missing in Action |
| ATT | Attack | M/W | Motor Works |
| BR | Bridge | M/Y | Marshalling Yard |
| C/I | Communications Installations | O/D | Ordnance Depot |
| C/T | Communications Target | O/I | Oil Installations |
| C/W | Chemical Works | O/R | Oil Refinery |
| Da | Damaged | O/S | Oil Storage |
| De | Destroyed | Pr | Probable Destroyed |
| DISP | Dispatched | R/J | Railroad Junction |
| E/A | Enemy Aircraft | RR/B | Railroad Bridge |
| G/D | Good Depot | RR/W | Railroad Works |
| G/I | Gun Installations | S/D | Submarine Docks |
| G/P | Gun Positions | TONS | Tons of bombs dropped |
| H/I | Harbor Installations | T/T | Tactical Targets |
| HW/B | Highway Bridge | T/W | Tank Works |

# CONCLUSION

After our last mission we were ordered to pack up because we were moving out. The war was over except for the formalities. Our's was the first Group in the Wing to break camp. Some of us were sent to the 464th Bomb Group where we celebrated VE Day. Others were sent to Naples to catch a ship home and reassignment. The lucky ones flew a B-24 back to the states.

We left Italy with mixed feelings. We were elated that we were going home. At the same time we were sad that so many of our buddies gave their lives for the cause. We also had a sense of guilt that we were spared while so many good men died. We still carried the scars in our memories. They would slowly fade; however, some will never fade

But we had completed our mission and were returning to a more normal life; going back to our jobs or continuing our education or marrying that special girl. We had helped preserve our freedom. However, it wasn't until years later, when the information was released, that we realized how unprepared we were to fight a war and how close we came to losing everything.

With the new weapons and methods of fighting we tend to think we will never lose our freedom, but we still must be ever watchful not only for the enemy from without, but also for the enemy from within.

*Lynn Colleman, navigator*

(Right) Sometimes you get lucky and the shell goes through the wing and explodes above the plane. When a shell explodes near your plane, the concussion rocks the old bird and the exploding shell peppers the plane with pieces of flak. (Note the smaller holes in the wing.) In this photo an aircraft is inspected after its safe return to Venosa.

(Below) *Mickey Finn*, an 828th Squadron aircraft, on the runway at Venosa, with a collapsed nose wheel.

# THE HIGH COSTS OF ACCIDENTS

**10 October 1944-"Red L" (44-41068) from the 828th Squadron burned at the base.**

**1 November 1944- 828th Squadron plane "Red M" (42-51274) crashed after take-off , killing all personnel aboard.**

**11 November 1944-829th Squadron plane 42-78299 crashed after take-off and seven airmen died in the crash, while three parachuted safely.**

# MISSION INDEX

## TARGET ABBREVIATIONS USED

| | | | | | |
|---|---|---|---|---|---|
| A/CF | Aircraft Factory | H/I | Harbor Installations | O/R | Oil Refinery |
| A/D | Air Drome | HW/B | Highway Bridge | O/S | Oil Storage |
| AM/D | Ammunition Dump | I/A | Industrial Area | PFF | Path Finder |
| BR | Bridge | LD | Locomotive Depot | RR/B | RR Bridge |
| C/I | Communications Install | M/W | Motor Works | RR/S | RR Station |
| C/W | Chemical Works | M/W | Motor Works RR/S | S/D | Sub Docks |
| E/W | Engine Works | M/Y | Marshalling Yards | T/T | Tactical Target |
| G/D | Goods Depot | O/D | Ordnance Depot | T/W | Tank Works |
| G/I | Gun Installations | O/I | Oil Installations | | |

| NO. | DATE 1944 | | TARGET | | PAGE NO. |
|---|---|---|---|---|---|
| 1. | 10 May | M/Y | Knin, Yugoslavia | | 31 |
| 2. | 12 May | M/Y | Via Reggio, Italy | | 31 |
| 3. | 13 May | M/Y | Modena, Italy | | 31 |
| 4. | 14 May | M/Y | Mestre, Italy | | 31 |
| 5. | 17 May | M/Y | Piombino, Italy | | 32 |
| 6. | 18 May | BR | Nis, Yugoslavia | | 32 |
| 7. | 19 May | M/Y | Bologna, Italy | | 32 |
| 8. | 22 May | M/Y | Valmontone, Italy | | 32 |
| 9. | 23 May | M/Y | Valmontone, Italy | | 32 |
| 10. | 24 May | A/D | Wiener-Neustadt, Austria | | 33 |
| 11. | 25 May | M/Y | Amberieu, France | | 33 |
| 12. | 26 May | M/Y | Lyon, France | | 33 |
| 13. | 27 May | M/Y | Nimes, France | | 33 |
| 14. | 29 May | A/CF | Vienna, Austria | | 34 |
| 15. | 30 May | A/CF | Neunkirchen, Austria | | 34 |
| 16. | 31 May | O/R | Ploesti, Rumania | | 34 |
| 17. | 2 June | M/Y | Cluj, Rumania | | 37 |
| 18. | 4 June | M/Y | Turin, Italy | | 37 |
| 19. | 5 June | M/Y | Forli, Italy | | 37 |
| 20. | 6 June | O/R | Ploesti, Rumania | | 38 |
| 21. | 7 June | H/I | Leghorn, Italy | | 48 |
| 22. | 9 June | A/CF | Munich, Germany | | 39 |
| 23. | 10 June | M/Y | Trieste, Italy | | 41 |

| NO. | DATE | | TARGET | | PAGE NO. |
|---|---|---|---|---|---|
| 24. | 11 | June | O/R | Smederevo, Yugoslavia | 42 |
| 25. | 13 | June | O/D | Munich, Germany | 42 |
| 26. | 14 | June | O/R | Peffurdo, Hungary | 43 |
| 27. | 16 | June | O/R | Vienna, Austria | 43 |
| 28. | 22 | June | M/Y | Castell Maggiore, Italy | 46 |
| 29. | 23 | June | O/I | Guirgui, Rumania | 46 |
| 30. | 25 | June | O/R | Sete, France | 47 |
| 31. | 26 | June | O/R | Vienna, Austria | 47 |
| 32. | 28 | June | O/R | Bucharest, Rumania | 48 |
| 33. | 30 | June | M/Y | Split, Yugoslavia | 50 |
| 34. | 2 | July | M/Y | Budapest, Hungary | 55 |
| 35. | 3 | July | M/Y | Timisora, Rumania | 55 |
| 36. | 5 | July | S/D | Toulon, France | 55 |
| 37. | 6 | July | O/S | Porto Marghera | 56 |
| 38. | 7 | July | O/R | Blechhammer | 57 |
| 39. | 8 | July | O/R | Vienna, Austria | 57 |
| 40. | 12 | July | M/Y | Nimes, France | 58 |
| 41. | 13 | July | O/S | Porto Marghera, Italy | 59 |
| 42. | 14 | July | M/Y | Mantua, Italy | 59 |
| 43. | 15 | July | O/R | Ploesti, Rumania | 60 |
| 44. | 16 | July | A/CF | Weiner-Neustadt, Austria | 60 |
| 45. | 19 | July | M/W | Munich, Germany | 60 |
| 46. | 20 | July | A/CF | Friedrichshafen, Germany | 61 |
| 47. | 22 | July | O/R | Ploesti, Rumania | 62 |
| 48. | 24 | July | A/D | Valonce, France | 65 |
| 49. | 25 | July | T/W | Linz, Austria | 65 |
| 50. | 26 | July | A/D | Vienna, Austria | 66 |
|  |  |  | A/D | Szombathley, Hungary | 66 |
| 51. | 28 | July | O/R | Ploesti, Rumania | 67 |
| 52. | 30 | July | A/D | Budapest, Hungary | 67 |
| 53. | 2 | Aug | H/I | Genoa, Italy | 71 |
| 54. | 3 | Aug | A/C | Friedrichshafen, Germany | 71 |
| 55. | 6 | Aug | RR/B | Tarascon, France | 72 |
| 56. | 7 | Aug | O/R | Blechhammer, Germany | 72 |
| 57. | 9 | Aug | A/D | Budapest, Hungary | 73 |
| 58. | 10 | Aug | O/R | Ploesti, Rumania | 74 |
| 59. | 12 | Aug | G/I | Sete, France | 75 |
| 60. | 13 | Aug | G/I | Sete, France | 75 |
| 61. | 14 | Aug | G/P | St. Tropex, France | 75 |
| 62. | 15 | Aug | HW/B | Pont St Esorit, France | 76 |
| 63. | 16 | Aug | C/W | Friedrichshafen, Germany | 76 |
| 64. | 18 | Aug | O/R | Ploesti, Rumania | 76 |
| 65. | 21 | Aug | A/D | Nis, Yugoslavia | 77 |
| 66. | 22 | Aug | O/S | Vienna, Austria | 77 |
| 67. | 23 | Aug | A/D | Markersdorf, Austria | 78 |

| NO. | DATE | | TARGET | | PAGE NO. |
|---|---|---|---|---|---|
| 68. | 24 | Aug | O/S | Pardubice, Czechoslovakia | 79 |
| 69. | 25 | Aug | A/D | Prostejov, Czechoslovakia | 80 |
| 70. | 27 | Aug | O/R | Blechhammer, Germany | 80 |
| 71. | 28 | Aug | O/R | Szony, Hungary | 81 |
| 72. | 29 | Aug | M/Y | Moravska/Ostrava, Czech. | 81 |
| 73. | 1 | Sept | M/Y | Szajol, Hungary | 86 |
| 74. | 4 | Sept | RR/B | Adige/Ora, Italy (Brenner Valley) | 86 |
| 75. | 5 | Sept | RR/B | Szob, Hungary | 86 |
| 76. | 6 | Sept | M/Y | Nyiregyhaza, Hungary | 87 |
| 77. | 10 | Sept | I/A | Vienna, Austria | 87 |
| 78. | 12 | Sept | C/CF | Wasserburg, Germany (Jet Factory) | 88 |
| 79. | 13 | Sept | O/R | Oswiecim, Poland | 88 |
| 80. | 17 | Sept | O/R | Budapest, Hungary | 89 |
| 81. | 18 | Sept | RR/B | Budapest, Hungary | 90 |
| 82. | 20 | Sept | M/Y | Hatvan, Hungary | 90 |
| 83. | 21 | Sept | M/Y | Brod, Yugoslavia | 91 |
| 84. | 22 | Sept | E/W | Munich, Germany | 91 |
| 85. | 23 | Sept | RR/B | Latisana & Piave, Italy | 92 |
| 86. | 24 | Sept | M/Y | Salonika, Germany | 93 |
| 87. | 4 | Oct | O/S | Vienna, Austria | 96 |
| 88. | 7 | Oct | O/S | Vienna, Austria | 96 |
| 89. | 10 | Oct | RR/B | Susegana, Italy | 97 |
| 90. | 12 | Oct | AM/D | Bologna, Italy | 97 |
| 91. | 13 | Oct | M/Y | Banhida, Hungary | 98 |
| 92. | 16 | Oct | A/CF | Graz, Austria | 98 |
| 93. | 17 | Oct | M/Y | Nagykaniza, Hungary | 99 |
| 94. | 20 | Oct | M/Y | Rosenheim, Germany | 99 |
| 95. | 23 | Oct | E/W | Augsburg, Germany | 100 |
| 96. | 1 | Nov | M/Y | Graz, Komend and | 102 |
| | | | | Studenzen, Austria | 102 |
| 97. | 3 | Nov | M/Y | Munich, Germany (3 planes PFF) | 102 |
| 98. | 4 | Nov | O/R | Linz, Austria | 103 |
| 99. | 4 | Nov | | Podgorica, Yugo (2 Planes) | 103 |
| 100. | 5 | Nov | O/R | Vienna, Austria | 103 |
| 101. | 5 | Nov | | Podgorica, Yugo (6 Planes) | 104 |
| 102. | 6 | Nov | O/D | Vienna, Austria | 104 |
| 103. | 7 | Nov | RR/B | Ora, Italy (Brenner Valley) | 105 |
| 104. | 15 | Nov | O/R | Linz, Austria (3 planes PFF) | 106 |
| 105. | 16 | Nov | M/Y | Munich, Germany | 106 |
| 106. | 17 | Nov | O/R | Blechhammer, Germany | 107 |
| 107. | 18 | Nov | A/D | Udine, Italy | 108 |
| 108. | 19 | Nov | A/D | Horsching, Austria | 108 |
| 109. | 20 | Nov | O/R | Blechhammer, Germany | 108 |
| 110. | 22 | Nov | M/Y | Salzburg, Austria | 109 |
| 111. | 25 | Nov | M/Y | Munich, Germany (1 Plane) | 110 |

| # | Date | Type | Target | Pg |
|---|---|---|---|---|
| 112. | 3 Dec | MN | Klagenfurt, Austria (2 Planes PFF) | 112 |
| 113. | 6 Dec | M/Y | Hegyeshalom, Hungary | 112 |
| 114. | 8 Dec | RR | Gliesdorf, Austria (RR Junction) | 112 |
| 115. | 12 Dec | O/R | Blechhammer, Germany (I Plane) | 114 |
| | | O/R | Moravska/Ostrava, Czech (1 Plane) | 114 |
| 116. | 15 Dec | MN | Salzburg, Austria | 114 |
| 117. | 16 Dec | O/I | Pilsen, Czech. | 115 |
| 118. | 17 Dec | O/R | Blechhammer, Germany | 115 |
| 119. | 18 Dec | O/R | Blechhammer, Germany | 116 |
| 120. | 19 Dec | MN | Maribor, Yugoslavia | 117 |
| 121. | 20 Dec | O/R | Brux, Czech | 118 |
| 122. | 26 Dec | O/R | Blechhammer, Germany | 118 |
| 123. | 27 Dec | MN | Maribor, Yugo; Villach & | 119 |
| | | | Klagenfurt, Austria | 119 |
| 124. | 28 Dec | O/R | Kralupy, Czech | 120 |
| 125. | 29 Dec | M/Y | Verona, Italy | 120 |

### 1945

| # | Date | Type | Target | Pg |
|---|---|---|---|---|
| 126. | 8 Jan | M/Y | Villach, Austria; Salzburg | 123 |
| 127. | 19 Jan | M/Y | Zagreb, Yugoslavia | 123 |
| 128. | 20 Jan | C/I | Linz, Austria | 124 |
| 129. | 31 Jan | O/R | Moosbierbaum, Austria | 124 |
| | | M/Y | Maribor, Yugoslavia | 124 |
| 130. | 1 Feb | | Graz, Austria | 127 |
| 131. | 5 Feb | O/S | Regensburg, Germany | 127 |
| 132. | 7 Feb | O/S | Pola, Italy | 131 |
| 133. | 8 Feb | G/D | Vienna, Austria | 131 |
| 134. | 9 Feb | O/R | Moosbierbaum, Germany | 132 |
| 135. | 13 Feb | M/Y | Graz, Austria | 132 |
| 136. | 13 Feb | G/D | Vienna, Austria | 134 |
| 137. | 14 Feb | O/R | Vienna, Austria | 134 |
| 138. | 15 Feb | RR/S | Wiener Neustadt, Austria | 135 |
| 139. | 16 Feb | A/D | Regensburg, Germany | 135 |
| 140. | 17 Feb | H/I | Pola, Italy | 142 |
| 141. | 19 Feb | H/I | Pola, Italy | 142 |
| 142. | 20 Feb | H/I | Trieste, Italy | 143 |
| 143. | 21 Feb | M/Y | Vienna, Austria | 143 |
| 144. | 22 Feb | | Casarsa, Italy | 144 |
| 145. | 23 Feb | | Bruck, Austria | 145 |
| 146. | 24 Feb | | Aborted | 146 |
| 147. | 25 Feb | O/R | Linz, Austria | 146 |
| 148. | 27 Feb | M/Y | Augsburg, Germany | 146 |
| 149. | 27 Feb | M/Y | Ora, Italy (Brenner Valley) | 147 |
| 150. | 1 Mar | M/Y | St Polen, Austria | 149 |
| | | M/Y | Amstetten, Austria | 149 |
| 151. | 2 Mar | M/Y | Linz, Austria | 149 |
| 152. | 4 Mar | M/Y | Szombathely, Hungary | 151 |
| 153. | 8 Mar | M/Y | Verona, Italy | 151 |

| NO. | DATE |  | TARGET |  | PAGE NO. |
|---|---|---|---|---|---|
| 154. | 9 | Mar | M/Y | Graz, Ausria | 152 |
| 155. | 12 | Mar | O/R | Vienna, Austria | 152 |
| 156. | 13 | Mar | M/Y | Regensburg, Germany | 153 |
| 157. | 14 | Mar | M/Y | Nove Samky, Hungary | 153 |
| 158. | 15 | Mar | M/Y | St Polten, Austria | 154 |
| 159. | 16 | Mar | M/Y | Amstetten, Austria | 155 |
| 160. | 19 | Mar | M/Y | Muhldorf, Germany | 155 |
| 161. | 20 | Mar | M/Y | Amstetten, Austria | 158 |
| 162. | 21 | Mar | A/C | Neuberg, Germany | 158 |
| 163. | 22 | Mar | M/Y | Vienna, Austria | 159 |
| 164. | 23 | Mar | M/Y | Gmund, Austria | 160 |
| 165. | 24 | Mar | A/D | Neuburg, Germany | 160 |
| 166. | 25 | Mar | A/D | Prague, Czech | 161 |
| 167. | 26 | Mar | M/Y | Bratislava, Czech | 161 |
|  |  |  | M/Y | Szombathely, Hungary | 161 |
| 168. | 30 | Mar | G/D | Vienna, Austria | 162 |
| 169. | 31 | Mar | M/Y | Villach, Austria | 163 |
| 170 | 1 | Apr |  | Bratislava, Czech (Aborted) | 165 |
| 171 | 2 | Apr | M/Y | Graz, Austria | 165 |
| 172 | 5 | Apr | L/D | Turin, Italy | 165 |
| 173 | 7 | Apr |  | Aborted | 168 |
| 174 | 8 | Apr | M/Y | Fortezza, Italy (Brenner Valley) | 168 |
| 175 | 9 | Apr | T/T | Front Lines, N. Italy | 169 |
| 176 | 10 | Apr | T/T | Front Lines, N. Italy | 169 |
| 177 | 11 | Apr | RR/B | Campo Di Trens, Italy | 169 |
| 178 | 12 | Apr | RR/B | Ponte Di Piave, Italy | 170 |
| 179 | 14 | Apr | MT/D | Ossopo, Italy | 171 |
|  |  |  | M/Y | Klagenfurt, Austria | 171 |
| 180 | 15 | Apr | G/I | Bologna Area, Italy | 171 |
| 181 | 16 | Apr | G/I | Bologna Area, Italy | 172 |
| 182 | 17 | Apr | G/I | Bologna Area, Italy | 172 |
| 183 | 19 | Apr | M/Y | Rosenheim, Germany | 173 |
|  |  |  | A/D | Udine, Italy | 173 |
| 184 | 20 | Apr | HW/B | Garzara, Italy | 173 |
| 185 | 23 | Apr | HW/B | Padua, Italy | 174 |
|  |  |  | HW/B | Cavarzer, Italy | 174 |
| 186 | 24 | Apr | HW/B | Casarse, Italy | 175 |
| 187 | 25 | Apr | M/Y | Linz, Austria | 176 |

www.ingramcontent.com/pod-product-compliance
Lightning Source LLC
Chambersburg PA
CBHW080544170426
43195CB00016B/2668